Savannah
DC
LA
Irvine

PERSONALITY, POLITICS, AND PLANNING

Personality, Politics, and Planning

How City Planners Work

Edited by
Anthony James Catanese
and **W. Paul Farmer**

HT
167
.P45
WEST

 SAGE PUBLICATIONS   Beverly Hills / London

Copyright © 1978 by Sage Publications, Inc.

All rights reserved. No part of this book may be reproduced or utilized in any form or by any means, electronic or mechanical, including photocopying, recording, or by any information storage and retrieval system, without permission in writing from the publisher.

For information address:

SAGE PUBLICATIONS, INC.
275 South Beverly Drive
Beverly Hills, California 90212

SAGE PUBLICATIONS LTD
28 Banner Street
London EC1Y 8QE

Printed in the United States of America

Library of Congress Cataloging in Publication Data

Main entry under title:

Personality, politics, and planning.

    Includes index.
    1. City planning—United States.
I. Catanese, Anthony James.   II. Farmer, W. Paul
HT167.P45   309.2'62'0973   77-17780
ISBN 0-8039-0961-6

FIRST PRINTING

CONTENTS

Chapter		Page
	Acknowledgment	7
	Preface	9
1.	Framework for Comparison by W. Paul Farmer	11
2.	Atlanta: Planning, Budgetting, and Neighborhoods by Leon S. Eplan	33
3.	Cleveland: Problems of Declining Cities by Norman Krumholz	57
4.	Indianapolis: Fragmentation and Consolidation by Michael A. Carroll	79
5.	Kansas City: Problems and Successes of Downtown Development by Joseph E. Vitt, Jr.	103
6.	Milwaukee: Planning for Fiscal Balance by William Ryan Drew	131
7.	Portland: The Problems and Promise of Growth by Ernest Bonner	145
8.	St. Paul: Professionalization and Timing of Planning by Donald L. Spaid	159
9.	Learning by Comparison: Lessons from Experiences by Anthony James Catanese	179
	Appendix	209
	Index	223
	About the Editors	227

ACKNOWLEDGMENT

We would like to acknowledge the support and assistance of our colleagues, Dr. David S. Sawicki, Dr. Ernest Alexander, Dr. Joseph Mangiamele, Dr. Sammis White, Dr. James Snyder, Dr. Larry Witzling, and Professor David Hoeh. The greatest thanks go to Ms. Mary Eichstaedt for her devotion above and beyond the call of duty in the typing, proofreading, and production of this book.

A.J.C.
W.P.F.

Milwaukee, Wisconsin
April 1977

PREFACE

These talks by the chief planners of seven of our largest central cities represent an effort to compare the way that city planning takes place in America today. We initially sought to compare and examine: (1) the theories of planning in use, (2) innovations and new ways of doing planning, (3) major successes and apparent failures, and (4) what lessons could be learned from these experiences that could be useful to other cities. All of these topics were posed as questions included in our instructions to the city planners. We quickly found, however, that there was a human perspective that transcended theory, practice, and implementation. The personalities, styles, and philosophies of these planners seem to be equally significant in determining how one does planning— and survives. We also discovered that comparative analysis of city planning is difficult in such a structured approach and necessitates wider areas of examination.

These conversations were sponsored by the Department of Urban Planning, School of Architecture and Urban Planning, University of Wisconsin—Milwaukee. They took place over a four-month period from September to December 1976 and were edited and revised in January 1977.

We hope that this book will help fill what we consider to be a major gap in our knowledge of contemporary city planning experience. We believe that the comparative, case approach is vital for learning more about planning and implementation. This book may very well be the first of its kind—we sincerely hope that more will follow.

Chapter 1

FRAMEWORK FOR COMPARISON

W. PAUL FARMER

Assistant Professor of Urban Planning
School of Architecture and Urban Planning
University of Wisconsin—Milwaukee

INTRODUCTION

The personalities, professional training, and values of the city planners whose talks are included in this book are all variables which influence the planning activities and planning processes of the seven cities.

This chapter examines only basic economic and population data of the seven cities in order to provide a basis of understanding for the talks which follow. The cities are described with respect to population, employment, income, housing, education, transportation, federal programs, and overall quality of life. All tables referred to in this discussion may be found in the Appendix.

POPULATION

The image of the declining Northeast and Midwest and the rise of the cities of the West and South—the so-called Sunbelt—is by now well established in the popular press and news magazines. These processes of regional decline and growth are being studied seriously by professionals and academics concerned with the field of urban and regional planning. The seven cities in these talks provide a basis for examination of the specific processes which are ongoing in several regions of the United States.

Table 1 gives some indication of the growth that was being experienced by the manufacturing cities of the Northeast and the Midwest during the early decades of this century. The great industrial expansion and European ethnic immigration of the early decades of this century resulted in tremendous growth of these manufacturing centers. Cleveland, for example, grew by almost 50% from 1900 to 1910 and from 1910 until 1920 added an additional 230,000 residents, an increase of 42%. Kansas City grew by comparable amounts from 1900 until 1920, and its sizable growth rate of 23% extended into the 1920s. Indianapolis likewise experienced substantial growth into the depression years of the 1930s. The growth of these cities mirrors the industrial growth of the Northeast where the manufacturing plants of the early industrial revolution had located. It is significant that Cleveland's population was the largest of the seven cities by 1900, reaching almost 400,000, and that its population growth had slowed by 1930 and had been reversed by 1940. By 1970 the City of Cleveland had a smaller population than it had in 1920. The expansion of the industrial cities westward is indicated by Indianapolis and Kansas City, which maintained rapid urban growth into the 1930s. Indianapolis has continued that urban growth to the present time, although its increase of over 50% from 1960 to 1970 was due to its consolidation with Marion County.

Atlanta, the only southern city in these talks, is not only one of the South's oldest cities, but also has been one of the South's most rapidly growing cities since World War II. Atlanta's population growth, which was substantial in the early decades of this century, slowed during the 1930s and 1940s but then increased dramatically between 1950 and 1960 when the city population grew from 330,000 to almost one-half million. The decade from 1960 to 1970 was one of slow population growth for the City of Atlanta, although the metropolitan area continued to be one of the fastest growing in the country. These two aspects of Atlanta's population—early growth and, most recently, stagnation—suggest that Atlanta is not really one of the new cities of the United States, such as Phoenix, but rather has more in common with many of the older cities of the Northeast and Midwest, which face continued competition from their suburbs.

Portland may not be a prototypical city of the west coast as indicated by its population figures. Portland's most recent period of rapid growth occurred between 1940 and 1950, spurred by shipbuilding and related activities during World War II and by increased Pacific trade during the latter years of the decade. During the two most recent decades Portland has exhibited little of the growth usually associated with West coast cities.

It should be emphasized that the decline of central cities is a recent phenomenon occurring after the end of World War II. As noted by George Peterson in a study for the Urban Institute (p. 44):

> The image of the "declining" central city has become so firmly embedded in discussions of urban problems that it may be forgotten how recent a development actual population loss is. Of the sixty cities that had populations of 200,000 or more in 1960, only 38 percent declined in size between 1950 and 1960; in the sixties the proportion of such cities losing population rose to 45 percent; by 1970-73, it had reached 73 percent.

An even more recent phenomenon is the loss of population by entire metropolitan areas. From 1960 to 1970 the only large metropolitan area in the country to lose population was Pittsburgh. However, Table 2 indicates that from 1970-1973, the Cleveland metropolitan area lost almost 1% of its population each year; this contrasts with the previous decade during which it gained population at the rate of almost 1% per year. Although the other six metropolitan areas in these talks are still gaining population, according to the data through 1973, several of the areas have seen their population growth decline substantially from previous years. The Indianapolis metropolitan area grew by only slightly more than 25,000 people from 1970 until 1973, an annual percentage growth during that time of less than 1%. Kansas City experienced an annual growth from 1970 to 1973 of just slightly more than 1% or approximately 15,000 individuals per year. The Milwaukee and Portland metropolitan areas also experienced growth rates in the 1% to 2% per year range. Only the Atlanta and Twin Cities metropolitan areas experienced increases in their rates of growth from 1970 to 1973 compared to the previous decade. The Twin Cities saw an annual growth rate increase from slightly more than 2% to approximately 3.5% per year, while the Atlanta metropolitan area experienced a substantial increase from 3.7% per year during the 1960 to 1970 decade to 8.6% per year in the 1970 to 1973 period. Compared with the figures for Atlanta in Table 3, it is obvious that almost all of the growth in Atlanta from 1960 to 1970 occurred outside of the central city; that is true as well for the growth since 1970.

All seven of the metropolitan areas included in these talks fall into the one to three million population category, but they do not follow precisely the regional patterns of recent population growth and decline suggested for cities of that size by Vincent Barabba in a recent study. Cleveland has joined four other metropolitan areas—New York, Detroit, Pittsburgh, and Cincinnati—in terms of metropolitan decline, with all five areas located in the Northeast and Northcentral regions. The other cities in our talks do not support the regional patterns documented by Barabba—reduced southern growth rates and large decline in western growth rates, as well as absolute decline or stagnation in the Northeast and Northcentral areas. The population figures reported by Barabba do, however, suggest the broad regional patterns in the United States. These regional patterns certainly have great implications even for those metropolitan areas that have shown continued recent growth in the face of regional stagnation or decline, such as the Minneapolis-St. Paul metropolitan area compared with the Northcentral area.

Over the last fifty years the black population has played a major role in the changing patterns of population distribution among regions, states, and metropolitan areas. Table 4 shows the black population for the metropolitan areas and their central cities in 1970 and supports the frequent observation that the percentage of black population in U.S. cities is generally related both to the distance of cities from the southern states and to the manufacturing activity of cities. The Atlanta metropolitan area has a 22% black population, while Cleveland, one of the early destination cities of black rural-to-urban migration, is 16% black. Indianapolis and Kansas City, two cities which did not experience the early growth in manufacturing employment as did Cleveland, follow with 12% each of their population comprised of blacks. Milwaukee, which did not have a sizable black population until after World War II, has an 8% black population, while Portland and St. Paul still have black populations under 50,000, comprising 2% of their metropolitan area populations. The black population in this country is now an urban rather than a rural population and a central city rather than a suburban population. Table 4 demonstrates that in each of the seven cities the black population of the central city is at least two to three times the black population of the metropolitan area as a whole. In Milwaukee, for example, the city has a black population of 105,000, while the remainder of the metropolitan area has a black population slightly in excess of 1,000 individuals. While the City of Milwaukee in 1970 contained roughly 50% of the total population of the metropolitan area, it contained 99% of the black population of the metropolitan area.

While the black population group has had a major impact since World War II, the population groups that had such major impacts in the late nineteenth and first half of the twentieth centuries were largely foreign-born. These groups tended to provide the cheap labor required for manufacturing plant expansions during that period and in return were provided with both employment at appropriate skill levels and the necessary income and opportunity in the new world. The cities of the Northeast and later the Northcentral areas acted as the first recipients of these European immigrants. Just as Table 4 indicated the expansion of the black population from the South into other areas of the country, Table 5 indicates the expansion of the European immigrant population from the Northeast to the Industrial Belt stretching from the Atlantic to Lake Michigan. The figures for Cleveland and Milwaukee clearly indicate the role played in the development of those cities by the foreign-born—largely German, Polish, and Italian immigrants. The cities of the West coast have traditionally been the recipients of the immigrants who crossed the Pacific, as shown by the foreign-born population of Portland. Even though the Minneapolis-St. Paul area is well known for its Scandinavian heritage, Table 5 indicates that the immigration occurred at an earlier stage of city growth.

EMPLOYMENT

As previously discussed, Cleveland and Milwaukee are representative of the Industrial Belt cities that were the recipients of the European immigrant population during the late nineteenth and early twentieth centuries. This heritage is evident in Table 6, which shows the percentages of the employed population in manufacturing and white-collar activities for the seven cities. The distribution ranges from over 37% in Cleveland and almost 35% in Milwaukee to a low of only 17% in Portland and Atlanta. Atlanta, centrally located in the South and serviced by one of the major U.S. airports and by several interstate highway systems, is not a manufacturing city at all but rather a city that serves distribution purposes. Portland, St. Paul, Kansas City, and Indianapolis have all taken advantage of the continuing shift of our economy to service activities. Fifty-five percent of all employed persons in Portland are classified as holding white-collar jobs, while only 36% of Cleveland's population are so classified.

Cleveland provides an example of the differences between the central city population and the population of the remainder of the metropolitan area. While Cleveland's city population shows only 36% occupying white-collar jobs, 50% of the metropolitan area population is so classified. The percentage of the suburban population classified as white-collar is actually twice that of the City of Cleveland. Atlanta, too, experiences a wide variation between its central city and suburban populations in terms of white-collar employment.

Because the type of employment relates to the level of income for individuals and populations, it is an important indicator of economic health. An additional indicator is the unemployment rate. Table 7 shows that both the total population and the black population in the central cities of most of our seven urban areas experience a greater percentage of employment deprivation than do the suburban areas. The only exception to this relative deprivation appears in the 1970 Kansas City statistics for the black population. However, Kansas City is actually comprised of two central cities, and the metropolitan area figure contains both the suburban areas and the Kansas City, Kansas, central city. Table 7 also indicates that although the City/SMSA unemployment differences are substantial in some areas, an even greater difference is represented by the figures for the total population versus the unemployment rates for the black population in each of the metropolitan areas. There is not a single observation in which the black population has a lower unemployment rate than the total population; in fact, the unemployment rate for the black population in 1970 is frequently more than twice that for the nonblack population, whether looking at city figures or metropolitan figures.

The regional variations in unemployment rates are also evident, ranging from highs of over 5% in Cleveland as expected, and of over 6% in Portland, which might be unexpected, to lows of under 4% in Kansas City and St. Paul. The lowest metropolitan unemployment rate was in the rapidly growing Atlanta area.

INCOME

Unemployment and underemployment figures are important in determining the size of the population that has been left out of the overall system of economic progress. Occupational classifications also indicate relative ranks and statuses of individuals and the distributions of these throughout labor market areas. Median family income is one indicator that combines the other indicators of quantity and quality of the labor experience, as seen in Table 8. In general the ranking of regions by median income has remained constant over time, with northeastern and western areas showing the highest income and northcentral and southern areas showing the lowest incomes. While the rank of the regions remains the same, however, the regional differences have been narrowing over time. Obviously, within regions some cities could be considered to be relatively rich while others could be considered relatively poor. Two of the cities in these talks exhibit such patterns. The median income, unadjusted for cost of living, of the Atlanta metropolitan area ranks as the fourth highest among the seven cities. It ranks ahead of both the Cleveland and the Kansas City metropolitan areas, even though both Cleveland and Kansas City are in regions of the country that have higher median incomes than the South. Milwaukee and Minneapolis-St. Paul had the highest median incomes in 1970. The former is a stable unionized manufacturing city, while the latter is a more rapidly growing white-collar area and a distribution center for the region of the country west of the Great Lakes.

Although regional variations in median income are narrowing, the differences between cities and their suburbs are frequently increasing. The city/metropolitan area ratios of the seven cities in these talks range from lows of 79% and 89% in Atlanta and Cleveland respectively, to a high of 100% in Indianapolis. This Indianapolis figure is obviously affected by the incorporation of many formerly suburban areas into the city definition due to the Indianapolis/Marion County consolidation which occurred prior to the 1970 census. Kansas City and St. Paul, each with smaller percentages of their metropolitan area populations than is the case in Indianapolis, both had median incomes of 94% of the metropolitan median. Milwaukee and St. Paul had median incomes of 91% of their metropolitan area medians. These city/suburban differences exist even in Portland with a small minority population, and the findings are consistent with studies that have found that city/suburban differences are still significant even after correcting for a larger central city proportion of black population and female heads of household, groups with generally lower incomes than white male heads of households.

The median income figures do not, however, reveal the actual income distributions for the cities and metropolitan areas. Table 9 shows the concentration of families at both the bottom and top of the income ladder. It shows that in every case the central city has a higher level of poverty, as indicated by the greater percentage of families with income below the poverty level, and a lower degree of affluence, indicated by lower percentages of families with income $15,000 or above. These differences range from Cleveland, with a city poverty population in 1970 of almost 14% compared with less than 7% for the metropolitan area to Indianapolis, with only a marginal difference of 7% for the city versus 6.5% for the metropolitan area. The incidence of affluent families ranges from Cleveland's low of 15% to Indianapolis' high of almost 25%. Again the city/suburb discrepancy is greatest for Cleveland and is narrowest for Indianapolis. Milwaukee, St. Paul, and Atlanta all show major city/metropolitan area differences in terms of both affluence and poverty. However, the city/metropolitan differences for poverty and affluence in both

Kansas City and Portland are not outstanding.

The greater extent of poverty exhibited in Atlanta and Cleveland is particularly striking given the major programs throughout the decade of the 1960s aimed at eradication of poverty. These programs, commonly known as the War on Poverty, followed the riots that began in 1964. They included programs such as the 1964 Economic Opportunity Act and Model Cities Act of 1966. These acts funded community organization efforts, educational programs, manpower training programs, health programs, and legal services programs. Although recent evaluations of these programs have indicated that they were more successful than initially believed, certainly these 1970 figures indicate that urban poverty remains a major problem for city planners and administrators. These data support the generalization of the suburbanization of high-income whites and centralization of low-income blacks.

HOUSING

The relative affluence shown in Tables 8 and 9 is translated into purchasing power as shown by the figures for median house values seen in Table 10. The distinction between the metropolitan housing market and city submarkets is also evident in Table 10 with the greatest difference in values again being found in Cleveland, where houses in 1970 were valued at less than $17,000 in the central city and at almost $23,000 in the metropolitan area. The smallest difference was in Indianapolis, where houses in the city were valued at only $100 less than the median for the metropolitan area. While these figures demonstrate again the plight of the central city, they also mask the fact that a metropolitan area housing market is actually made up of many submarkets in towns, villages, and neighborhoods. Aggregate house value is based upon many factors—construction cost, geographical location, relative growth of the metropolitan area, and the like. Since southern cities usually have housing values that are less than those of midwestern and western cities, it is interesting to note that Atlanta's housing values are substantially above those for Indianapolis, Kansas City, and Portland. Atlanta's rapid growth probably has much to do with the 1970 house values, which are higher than those that would be expected based on regional comparisons.

Nationally, new house prices have more than doubled in the last ten years, while interest rates have also increased. Housing prices and values have generally increased more rapidly than the consumer price index as a whole. For the first time in several decades the typical family's housing expenditure is now taking an increasingly larger percentage of the family income.

Owner occupancy is frequently considered an indicator of area affluence and stability. Nationally the percentage of owner-occupied dwelling units increased from 50% in 1950 to almost 60% in 1970. Owner occupancy in metropolitan areas of the seven areas in these talks ranges from a low of 57% in Atlanta to a high of 65% in Indianapolis, Minneapolis-St. Paul, Kansas City, and Portland, as seen in Table 11. The Atlanta metropolitan area is the only one in the study which exhibits a lower degree of home ownership than is the case nationally. However, each of the central cities exhibits a lower rate, with Atlanta's rate at slightly over 41% being the lowest. Since ownership of a home is one of the major forms of financial investment for individuals and families in the United States, and as investment has traditionally played that role as ethnic groups were able to advance upward, it is important to note that a large percentage of the population in the central cities of Atlanta, Cleveland, and Milwaukee did not have the opportunity for this type of savings investment. Lack of home ownership frequently makes neighborhood revitalization and reinvestment a more difficult task and often provides the basis for more conflicts between landlords and tenants.

Currently the standard indicator of housing need is the

degree of overcrowding, defined in the census as 1.01 or more persons per room, as shown in Table 12. In 1970 this degree of overcrowding in our seven cities ranged from a high of 11% in Atlanta to a low of 3.5% in Portland. The differences among the metropolitan areas were not as great, ranging from a high of 7.9% in Indianapolis to a low of 4.4% in Portland. Owner occupancy and degree of overcrowding are frequently not directly related as evidenced by a comparison of Tables 11 and 12. Under certain conditions it may be possible for individuals and families to make the choice between home ownership and a larger housing unit. Local conditions govern whether or not this choice is actually available.

The figures for overcrowding and owner occupancy, which indicate that Atlanta is experiencing significant housing problems, are probably related to the high black percentage of the population in Atlanta (Table 4). It can be hypothesized that the population is comprised of large black families, people who are relatively recent migrants to Atlanta and who have not been able to accumulate the savings necessary for a home purchase. Possibly, they have run into forms of discrimination which have discouraged home ownership and limited housing choice, but this is not clear from the data.

The year that structures were built is another indicator of housing need, as well as an indicator of the demand for revitalization and rehabilitation, as seen in Table 13. The earlier characterization of Cleveland as a city which was developed primarily during the late nineteenth and early twentieth centuries through the combination of industrialization and European immigration is supported by the figures in Table 13. These show that almost three-fourths of all dwelling units in the City of Cleveland were constructed prior to 1939, while only a thousand units, comprising less than 0.1% of the Cleveland housing stock in 1970, were built from 1969 to 1970. At the opposite extreme are the cities of Atlanta and Indianapolis with 30% and 40%, respectively, of their structures constructed prior to 1939 and 2.7% and 3.9%

built during the 1969 to 1970 period. Atlanta's image as an expanding city is certainly supported by the figures for the metropolitan area which show that less than 20% of all structures in the Atlanta area were constructed prior to 1939, while over 6% were constructed in a one year period—1969 to 1970. The Minneapolis-St. Paul metropolitan area also experienced rapid growth in housing units in the 1969-1970 period with almost 30,000 units being constructed, or slightly less than 5% of the overall housing stock of the Twin Cities area. Kansas City, Milwaukee, and Portland all fall into a middle range when comparing the seven cities of these talks. In these cities housing constructed prior to 1939 represents slightly more than half of the city housing stock and slightly less than half of the metropolitan area stock. All three cities also demonstrate relatively slow growth in housing stock as measured by units constructed in the 1969-1970 period.

These figures generally support the images projected by the seven cities in our discussions. Atlanta and the Twin Cities are perceived as being cities of growth and vitality, while Cleveland and, to a lesser extent, the other midwestern cities are perceived to be cities of decline or relatively stable conditions. It may be that the relative newness of housing in metropolitan areas is one of the more visible indicators of city conditions, and it is an indicator which seems to correlate quite closely with what might be termed "popular image." Certainly the figures for age of housing are consistent with such popularly conceived images, while many of the other indicators examined—median income, owner occupancy, and overcrowding—are not so consistent. These figures are consistent with the regional variations in housing construction, which have seen high increases in structures in the South and a much lower rate of increase in the North. In fact, many northcentral cities have seen a decline, while such decline in structures is very rare in southern metropolitan areas, as shown by Vincent Barabba (p. 65). who found that from 1960 to 1970 only ten of 106 central cities in the

South had experienced an absolute decline in housing stock. The age of housing is a particularly important housing indicator due to changing tastes and rises in real income, which result in newer units usually being larger with more amenities such as extra bathrooms. Thus a larger proportion of the housing stock in the South is likely to appeal to the typical family more than is the case for those housing units in the North.

The construction cost differentials between the North and South play an important role in the type of housing package that is actually available, as implied by Table 13. In 1973 the average price of new housing in the South was $33,200 while in the North the figure was $40,600. The price per square foot in the South has been approximately 15% less than the per square foot cost in the North for the last five years. Coupled with lower operating costs (due primarily to lower fuel costs and to less fuel needed for home heating), it is no surprise that the population movement noted earlier has been one primarily to Sunbelt cities of the South and West. Along with reasons such as lower labor costs and lower degree of unionization, the same construction cost figures and energy cost figures are reasons for the movement of industries (with their jobs) to the southern areas.

EDUCATION

Economic activities are attracted to cities partly by the quality of the labor force. Table 14 indicates that the labor forces in the seven cities vary widely along one dimension—the percentage of the population that had completed four years or more of high school in 1970. This one measure ignores the quality of education, effects of racial mix and segregation practices, and other possible indicators such as median school year completed and percentage of the population with college degrees. There are substantial differences among the cities with respect to completion of a high school education. Only slightly more than one-third of Cleveland residents over 25 hold high school diplomas, while the corresponding figure for residents of the City of Portland approaches two-thirds. In both Atlanta and Milwaukee less than 50% completed high school, while in Indianapolis, Kansas City, and St. Paul the figures are in the 55% to 57% range.

The educational differences among cities are not so pronounced for metropolitan areas, with all having more than half of the populations with high school degrees. Atlanta and Cleveland have the lowest percentages—53% and 54%—while Kansas City, Portland, and Minneapolis-St. Paul are the only metropolitan areas with greater than 60% of their populations with at least a high school education. The city/suburb distinction is again obvious from Table 14 as the metropolitan percentages are higher than the central city percentages in all seven cases. In Cleveland the difference is quite substantial.

TRANSPORTATION

The affluence of the population, modes of transportation, and physical forms of cities historically have been related. The dense mercantile cities of the eighteenth and nineteenth centuries in the United States gave way to the larger industrial cities of the late nineteenth and early twentieth centuries—cities in which densities first increased as technology allowed multistory construction and then decreased as the trolley-car, railroad, and automobile allowed decentralization. The land area required for residential, commercial, and industrial activities has increased over time so that density is highly correlated with the age of the city or the time periods of major development. Higher densities are both supportive of mass transportation and are supported by high service levels of such a mode of transportation. In contrast, the rate of automobile ownership relates to the affluence of the population while allowing and encouraging a low-density pattern of development.

Tables 15 and 16, showing land area and density and work trips by mode of transportation, clearly demonstrate these relationships, particularly when compared with Tables 6, 8, and 13, showing manufacturing employment, median income, and age of housing. The older industrial cities of Cleveland and Milwaukee have high densities, almost 10,000 and 8,000 persons per square mile, respectively, and high percentages of work trips dependent on public transportation. Atlanta, an older city, is in the middle density range (almost 4,000 persons per square mile) but shows a dependence on public transportation that almost matches that of Cleveland. This can be explained to some extent by the Atlanta income level, which is the lowest of the seven cities. St. Paul also has a high reliance on public transportation relative to cities such as Indianapolis, Kansas City, and Portland. St. Paul is an older city with a density of over 7,000 persons per square mile, approaching the density of Milwaukee. The physical features of Portland, which constrain decentralization, are a factor in its density (over 4,000 persons per square mile), although the lack of manufacturing activity and relatively high income level are factors that have reduced the transit ridership to lower than what might be expected based on density alone.

The decreasing densities from the central cities are evident for all seven cities, with the high central densities of Cleveland, Milwaukee, and St. Paul falling rapidly as measured by the densities recorded for the remainders of the urbanized areas and for the remainders of the metropolitan areas. Density does not fall as rapidly for those cities with lower central city densities.

Decentralization of activity in U.S. urban areas is not a recent phenomenon, nor is the decentralization of that activity across municipal boundaries. This latter process is commonly called "suburbanization," and has obvious negative effects on the central cities and their tax bases. The balkanization of city and suburb is a theme throughout these talks; further evidence of this process is shown in Table 17, which indicates the changes that occurred in workplace location during the decade of the 1960s. Recent evidence indicates that the suburbanization trend evident in Table 17 has accelerated in many cities during the 1970s. In all seven cities, the increases in jobs outside the central cities exceeded the central city increases by tremendous amounts. Cleveland, Milwaukee, and Kansas City actually suffered declines, while the areas outside the cities in those metropolitan areas had increases ranging from 56% to 110%. Three metropolitan areas—Milwaukee, Atlanta, and Indianapolis—had increases in the suburbs in excess of 100%, with the Indianapolis increase actually exceeding 200%. Workplaces located outside the central cities are usually dependent on the automobile for private employee transportation and can be served by traditional modes of mass transportation only with direct public subsidies in excess of those sometimes provided to urban systems.

FEDERAL PROGRAMS

The major federal program affecting planning and development activities in cities is currently the Housing and Community Development Act of 1974 (Public Law 93-383), which combined a number of programs into one block grant program as part of the Special Revenue Sharing Program. Seven major programs were consolidated: urban renewal, model cities, water and sewer facilities, open spaces, neighborhood facilities, rehabilitation loans, and public facility loans.

The Special Revenue Sharing Program represents a major departure from previous urban programs sponsored by the federal government, which had essentially invited communities to compete for special purpose funds with federal agencies setting detailed requirements and deciding allocations of funds. Under the Community Development Block Grant Program (CDBG) a large proportion of the funds is awarded to units of government on an "entitlement" basis, with actual dollar amounts based on a formula that takes into account population, poverty, and overcrowded housing. An additional safeguard, the "hold-harmless" clause, guarantees units of local government that they will receive funds during the first three years of the program which are at least equal to the average they received during 1968-1972 under the "folded-in" programs.

However, the hold-harmless guarantee declines until 1980 when it will be phased out completely and most funds will be allocated strictly on a formula basis. A study by the Brookings Institution indicates that central cities, such as those included in our talks, will lose money under the CDBG Program in 1980 compared to what they could have expected under the pre-1974 programs. As noted by Richard Nathan (p. 151):

> The projections to 1980 indicate losses in the aggregate [to central cities] of more than $300 million, with the central cities' share of the total funds declining from 71.8 percent to 42.2 percent .... The central cities are the only jurisdictions within metropolitan areas projected to lose funds under the formula allocation system compared to the hold-harmless base.

The actual losses projected for the cities in our talks range from slightly over 11% in Cleveland to more than 50% in Kansas City, as shown in Table 18.

The pre-1974 system of allocations rewarded cities for their definition of problems in light of federal programs and for their aggressive pursuit of federal dollars to aid in solving these problems. These cities stand to lose funds under the current formulas, which have been developed in order to meet the main objective of the CDBG Act—"the creation of viable urban communities." The Brookings Study questions the use of indicators of population, overcrowded housing, and poverty as the best measures of community need and concludes (Richard Nathan, p. 187):

What is missing from the formula is some specific measures [sic] of physical need to serve as an index of the condition of a community's physical environment—such as streets, curbs, sewers, as well as actual dwellings. For this aspect of community development need, one statistical indicator in the 1970 census that can be used for this purpose is the amount of housing stock built prior to 1939. Not all housing built before 1939 is deteriorated or deteriorating, but the age of housing is, in our view, quite clearly linked to the rehabilitation needs of urban communities and to the physical development purpose of the CDBG program.

The age-of-housing factor is particularly relevant to the needs of central cities, especially the neediest cities of the northeast quadrant.

Without an "age" factor in the formula, the CDBG Program results in funds flowing away from the older cities of the Northeast and Midwest to the newer Sunbelt cities of the South and West.

Due to a high correlation between two of the three factors currently used in the formula—extent of poverty and overcrowded housing—and due to the lack of such correlation between poverty and pre-1939 housing, the authors of the Brookings study recommended that the allocation formula be changed to benefit the "neediest" cities. Such hardship cities are characterized by "old age, an increasing concentration of the socially and economically disadvantaged, and population decline" (Richard Nathan, p. 509). A formula using the extent of poverty, population change, and pre-1939 housing results in an eligibility index that shows six of our seven cities above the national mean. For example, Cleveland would rank eighteenth nationally according to this eligibility index, according to Table 19. Previous tables have indicated that Cleveland is characterized by population decline, substantial poverty, and a very high proportion of housing stock constructed prior to 1939. Five of the cities in our talks cluster around the 134-164 rankings, while Indianapolis is below the national mean for the eligibility index. The inclusion of suburban areas, with their newer housing, in the 1970 definition of the city partially accounts for this ranking.

QUALITY OF LIFE

Just as agreement on a formula that evaluates relative need is likely to be elusive, agreement on a formula that evaluates relative quality of life is certain to be debatable. A perception of quality of life, however defined, is an important variable in an increasingly mobile society. As an ever smaller percentage of jobs is tied directly to natural resources (such as arable farm land and deep-mined coal) or to markets in the old population centers of the East, amenities become more important in locational decisions. Quality of life variables for most individuals are likely to include the following types of factors: economic, political, environmental, social, health, and educational.

In a study by Ben-Chieh Liu using these indicators for the U.S. Environmental Protection Agency, it was determined that the Portland metropolitan area ranked number one in the country in its quality of life. It achieved this high overall ranking by receiving the highest ranking, "A," in each of the five individual categories, as shown in Tables 20 and 21. The Minneapolis-St. Paul area was ranked number four out of the sixty-five largest metropolitan areas, while Atlanta received the lowest ranking of the cities in our talks. It was ranked forty-fifth, receiving an "A" rating only for the economic component.

While it can be argued that such variables as the local newspaper circulation per 1,000 population may not be a very good indicator to be included in determining the quality of life, the results in these tables seem to coincide with certain popular images, as well as with the data discussed in this chapter. The remainder of this book is concerned with the actions that can be taken by public agencies with respect to the conditions that have been examined. What are the planning activities being undertaken by Norman Krumholz in Cleveland, a city ranked eighteenth nationally on a hardship scale? What are the problems faced by Ernest Bonner in Portland, a city in "God's Country," ranked first nationally in quality of life? These and other questions are treated in some detail.

REFERENCES

George Peterson, "Finance," in William Gorham and Nathan Glazer (editors), *The Urban Predicament* (Washington, D.C.: The Urban Institute, 1976).

Vincent P. Barabba, "The National Setting: Regional Shifts, Metropolitan Decline, and Urban Decay," in George Sternlieb and James Hughes (editors), *Post-Industrial America: Metropolitan Decline and Inter-Regional Job Shifts* (New Brunswick, N.J.: Center for Urban Policy Research, 1975), pp. 59-64.

Richard P. Nathan, et al., *Block Grants for Community Development* (Washington, D.C.: U.S. Department of Housing and Urban Development, January, 1977).

Ben-Chieh Liu, *Quality of Life Indicators in U.S. Metropolitan Areas* (New York: Praeger, 1976).

Chapter 2

ATLANTA: PLANNING, BUDGETING, AND NEIGHBORHOODS

LEON S. EPLAN, AIP

Commissioner of Budget and Planning
Atlanta, Georgia,
and President, American Institute of Planners

We are struggling in Atlanta with the notion of decision-making. The way choices were given priority and made in the past no longer seems applicable. New forces have appeared which demand attention and which cannot be ignored. New faces, new methods, and new conditions are with us. I want to talk about what these new forces are and about Atlanta and how we are contending with these forces. Especially I would like to describe, first, how planning—specifically the comprehensive development plan—is evolving to become the central policy statement on how the city both develops and operates and, second, how citizens participate in this process.

Three years ago Atlanta was brought under a new City Charter. It was the first change in a hundred years, and it caused the city's form of government to change drastically. The Charter shifted administrative power from the Council to the Mayor by replacing a weak mayor with a strong mayor government. The Charter's companion, the Reorganization Ordinance, which set out the methods by which the strong executive was to function, consolidated 26 departments and agencies of government into nine departments. The new Charter provided for the election of City Council members by district rather than at large, a feature that has produced profound differences in the kinds of people who come to the Council. Finally, the Charter established a new policy-making, Office of Management and Budget-type bureau which is tied to the comprehensive plan.

The new Charter and Reorganization Ordinance which defined the form and function of the executive branch vary sharply from the old Charter in three important ways as related to comprehensive planning and budgeting. First, it establishes an organizational tie between urban planning and budget policy, placing these within a single agency. Second, it mandates a continuous comprehensive planning program. Third, it enormously increases citizen input into the process. I would like to take a minute to explain these three.

First the new *linkage between planning and budgeting*. Under the old Charter, enormous powers were vested in the Council. As in most strong council forms of government, that body had both legislative and executive functions. There were some two dozen Council Committees, most of which were composed of four or five members, and departments were assigned to these committees. There was a Finance Committee of the Council and a Finance Department, an Aviation Committee of the Council and an Aviation Department, and a Planning and Development Committee of the Council and a Planning Department. The committees, even more than the Mayor, were closely involved with the activities and priorities of the departments. Departments, in fact, served as staff to the committees. The department head and related council committee chairmen in most cases together fashioned policies and activities for the city in that particular area. Together they chose the programs, drew up the budget, and approved most of the activities. This gave the departments enormous independence. The problem was that under this system there was little or no coordination among the many departments with regard to programs, and there was no overall city policy.

When it came to running the city, the Mayor was little more than a figurehead. Atlanta had a succession of strong mayors—Bill Hartsfield (for 23 years), Ivan Allen, Sam Massell—but they were strong by virtue of their own personalities and not because of any powers that were vested in them. It should be of some interest that under that arrangement—without the executive institutionally capable of directing the policy of the city—it was the Finance Director who really had the power. The Finance Director had something of a negative power or a veto. He said that "this was how much money we had to spend," and when a request came in, he went through systematically and cut the budget back to that total. There is no deficit financing in Atlanta by law, so we have a very conservative financial policy. This hardline budget limitation gave considerable clout to the Finance Director when he began cutbacks. The policy was largely negative because it neither suggested programs nor outlined a policy framework against which cuts could be evaluated. There was, as a result, little programmatic relationship between one department's activities and those of another. There was little evaluation of the usefulness of programs and only the acceptance or veto of expenditures.

The new Charter cut this tie to the Council, condensed the agencies and departments, and turned department activities (which had focused on the Council) over to the Mayor. He is given the opportunities to establish the overall policy for the city and to direct all activities in a common direction. Development of that common direction takes place within the Department of Budget and Planning.

The old Planning Department has now been made into a Bureau called the Bureau of Budget Policy and Evaluation. The new bureau's duties are to recommend budgetary policy to the Mayor, formulate budget priorities, set computer priorities, evaluate programs, and monitor these consistently to goals. By placing these in the same department with planning, budget policies flow from the comprehensive development plan devised by the Bureau of Planning. The result has been that the budgetary process is closely linked to the comprehensive plan. So the first thing the Charter did was to build a close linkage between budget and planning.

The Charter did a second thing. *It mandated continuous, comprehensive, cooperative planning.* For the most part, the Charter is a very general document. It says what the Mayor should do in general terms, and then it is up to the Mayor and Council to work out the guidelines. In regard to planning, the Charter is more specific. Here the Charter makes three requirements. First, it says that each year the Mayor is to produce a comprehensive development plan for the city for a one-, five-, and fifteen-year period. Each year we produce a comprehensive plan. Second, the Charter mandates that, in the development of the annual plan, there must be substantial citizen involvement, and we have evolved a very elaborate process for doing that. Third, the Charter mandates that the Council will receive the Mayor's comprehensive plan, review it, hold additional hearings, and alter it if they like, but by law they must, no later than the middle of June, adopt that comprehensive plan. Because it is an annual plan, the Bureau of Planning begins the process of review and updating almost immediately by holding hearings and receiving input for the following year's plan. Therefore, that is where the notion of comprehensive, continuous planning is based. The procedure is strongly supportive of planning.

The Charter requires that the annual plans be produced and adopted by the Council, and, under this system, the one-year plan is particularly noteworthy. It becomes the first year of the capital improvements program. It is the annual zoning plan. We say at the beginning of the year what we want rezoned, and then we do not deviate very far except under extraordinary circumstances. If other rezonings are proposed, the Council is not simply reviewing a zoning petition, but it is actually deciding on a formal change in the legally adopted comprehensive development plan. This is a much more difficult chore to achieve, one that the Council is more reluctant to undertake. So we have comprehensive, continuous planning with a lot of clout.

The third innovation that the new Charter mandates is substantial *citizen involvement*. As has occurred in many other cities, citizens in Atlanta have become far more active and articulate than their predecessors. They have stopped several highways; they have successfully fought a number of rezoning questions; and they vigorously oppose any changes that affect their neighborhoods. While the response has been largely negative—that is, to stop change from occurring—some of this activity has been positive, such as their broad support in behalf of rapid transit. Citizen involvement was also felt in the adoption of the new Charter, and it was they, the citizens' groups around town, who sought the change to the Council District Election as a way of having their voices heard by the Council.

Let me back up for a minute and make some observations about the nature and history of citizen involvement in Atlanta. Atlanta is not typical of American cities. I would be hard pressed to name what I would consider a typical American city, as they are all quite a bit different from one another. I have to say Atlanta is perhaps representative of a fairly large number of newly emerging regional centers of national consequence; like many others, it has benefited and suffered from the trends of the post-World War II years. These trends are familiar: first, the incredible migration to a handful of metropolitan areas, in which almost 90% of our growth took place in 100 to 120 areas over the past thirty years, a phenomenon that drained the countryside and robbed huge portions of many states of many of their inhabitants. Second, the abandonment of central cities by the residents. People simply walked away from the institutions and the values they and their forebearers had built to build another city—a doughnut around the central city, creating the two Americas recognized by the National Commission on Civil Disorders. Then, finally, the habitation of the old central city by a new and different clientele plus those who remained behind (there are always in any great migration those who remain behind). This is a clientele composed of people who contrast considerably from those who left: people who are younger and poorer and blacker and less educated. Most believe that America has always been like it is today, divided into these two Americas. After all, this schism only began about thirty years ago, and more than half of the nation is under thirty years of age. But America before 1945 was significantly different and has always been different from what we see today. Trends of the last thirty years, in fact, have been what might be called an aberration—not normal— and I might add that this era, the last thirty-year period, is ending.

There are two important events occurring in many American central cities today that are challenging these trends of abandonment. One is the move to resettle older neighborhoods by young, middle class, largely white individuals and couples—a trend almost totally reversing the migrations of the past two or three decades. The other trend is the phenomenon of citizens seeking a greater and more active voice in the affairs of local government—again a reversal from the somewhat benign role citizens have played over the last several years. Both of these trends are taking place in Atlanta. We have a resettlement of the older areas and a greater demand for participation by the citizens. Atlanta is not unique; the same events are taking place in Pittsburgh, Portland, San Diego, Fort Worth, and Seattle, and to a lesser extent in Washington, Baltimore, and St. Louis. How it is occurring in Atlanta, and how the local government is responding, tell much about the way I believe central cities can survive and revive.

Neighborhoods have come alive in Atlanta! After several decades of silence and public apathy toward their well-being and decline, neighborhoods have become a major force in governmental decisions and spending priorities. Neighborhood advocates sit on the City Council and have the ear of the Mayor and executive branch. Their opinions are sought on all issues. This is a complete change from what was taking place only a few years ago. The country knows Atlanta as a vibrant city. Its remarkable growth during the 1950s and 1960s created a special spirit and image for it among American cities.

Perhaps Atlanta deserves that image, but to achieve this reputation was expensive. On the one hand, the changes produced enviable results; on the other, great violence was done to long-held values and established institutions. It was the best appearance as well as the worst. The 1960s brought prosperity to many. It gave Atlanta a national prosperity, a considerably wider variety of retail goods and services, improved cultural recognition, facilities in major league sports, and good restaurants. It was a swinging-singles style. But the swiftness and vastness of change also polluted our air, virtually lost for us our public transportation system, reduced the quality of public education, further separated the races, and allowed the level of public services, particularly streets and parks, to decline. The rewards were great, and the costs were very high.

In fact, we almost lost Atlanta during that period. In one ten-year period, between 1955 and 1965, due to public actions alone (primarily highway construction and urban renewal), we displaced 67,000 people. Fifteen percent of all Atlantans were moved out of their homes during that era, very few of them with public help. Commercial development, in addition, ate into the neighborhoods. Apartments replaced single-family dwellings. Highways ran through and divided and isolated communities, racial groups, and economic classes. Tens of thousands of dislocated people were simply shoved off the land and pushed into the adjacent neighborhoods, which were themselves fragile.

The programs launched by the City's Bureau of Planning are designed to attract people back to the city, to induce them to settle in these older inner-city neighborhoods, and to stabilize and rebuild them. We no longer believe, as we did in the 1960s, that governments can rebuild cities. We believe that people are the only city builders. The role of government today is far more passive, and our expectations are far more limited. We now encourage citizens to resettle these older areas, fix up the houses, return to the schools, and organize the churches. We are for government to get behind them, to help citizens where they cannot help themselves, to pave streets and repair sidewalks, to put in street lights and street furniture, and to protect and clean the area. This is not the sweeping role that government attempted to play in the 1960s, and it can only work where there is first a pioneering effort on the part of the citizens. It does recognize the current realities as well as the limitations, and it emphasizes the responsibilities of both government and citizens.

In Atlanta we have identified eleven neighborhoods which we refer to as "reviving." These are places where young people have moved back into the communities, reorganized themselves, and invested their personal time and energies, and they have begun to plan for their future. On the city's part, we have asked these and other neighborhoods to submit their plans for the future, to be included in the one-, five-, and fifteen-year comprehensive development plans, and to list improvements by neighborhood priorities, so that we can begin to hear what they have to say and to give priority to our programs based on their priorities.

One of the purposes of this exercise is to establish a certain degree of identity for neighborhoods. Atlanta has grown rapidly. The metropolitan area population jumped from 750,000 to 1,700,000 people in a period of only 25 years. During this period, as the region doubled in size, the city remained at about its current level of about 500,000 people. Since 1970, we have actually had somewhat of a decline. What we lost during this period of change was not simply houses and neighborhoods, but the bonds that tie people together: the garden clubs, boy scouts, PTAs, civic clubs, and church groups. This is the social fabric around which people build their collective lives and evolve themselves, and through which they share the responsibility for their fellow man. It was through these groups that we communicated with each other, and people were not only communicating with each other, but also with their government. When the dislocations of the 1950s and 1960s took place and these groups were broken up, people lost their ability to communicate with each other and with their government. When they lost that, they lost their faith in the government.

As the Atlanta region continues to grow—2 million people are expected by 1983, 3 million people by 1995—the inhabitants will find it increasingly difficult to relate to the vastness of the area. The infrastructure that exists, such as the twelve-lane highways and endless structures of commercial development, are hardly distinguishable one from another, especially for the young and old, for the housewives, for the infirm, for the mentally deficient, and for those raised in a less complex society. For perhaps a majority of the people of Atlanta, trying to understand the scale and the rate and intensity of change will not be easy.

So to make the metropolitan area understandable and less foreboding, we require that it be broken down into manageable parts, to a size a child can perceive and feel safe in. Such is the role of the neighborhood. The neighborhood is the small town within the large metropolitan area which, because of its human scale, can be understood, related to, and identified with. The planning efforts in Atlanta have been set to create a sense of identity between the people and the neighborhoods they live in.

In fact, we have found four major characteristics that we think are essential in creating a sense of identity in the neighborhood. First, like a child, a neighborhood needs a name. The requirement is for place recognition, so that a community can be referred to and endowed with pride and respect. In an otherwise large and anonymous urban region, the sense of belonging to some type of community grouping, preferably one that is geographically determined, is crucial. Also, a name often ties a person to a history, to a past, and, hence, to certain values and determinations of which way one is going. We have begun in Atlanta a series of neighborhood history projects so that people who move into an area can begin to get a sense of the history of that area and some understanding of what is around them and what things are of value. Second, in trying to establish these essentials of identity, we try to create a number of clearly defined boundaries. When one enters an area, one should know and feel that he or she is inside that area. When one leaves, one should know he or she is leaving. The medieval walled cities, for instance, created this kind of feeling. Of course, without the medieval walls in Atlanta, perhaps signs and columns and gates have to be used. The boundaries will aid the task of sharply separating the familiar from the unfamiliar. Third, there is a need for a central place within a neighborhood. This could be a location where people naturally gather; a building where meetings can be held; a park where children can drift in and find companionship; or a plaza or village center where food can be purchased or people can sit, meet, read, or simply stare at one another. At one time the schoolhouse served as a good central place, but that was before we lost our schools. Finally, in trying to create a sense of identity, there need to be organizations and recognized leaders. Organizations become important for sharing interests and concerns; they are a way for a person to relate to others. The leaders of such groups have become speakers and advocates on behalf of many. In Atlanta, we have increasingly turned to the neighborhood leaders for advice on issues affecting neighborhoods.

We identified some 190 neighborhoods in Atlanta, and we attempted to work with all of them. It was a failure. There were just too many individual neighborhoods to work with. They tended to balkanize the city, with one fighting another for whatever kind of aid and service they were seeking. To overcome that, we began to cluster the neighborhoods into larger units—units that had geographic interests or shared concerns but which were simply related to one another. We call them *Neighborhood Planning Units*. They are NPUs, so six or eight neighborhoods now make up an NPU. The average size of the NPU is something on the order of 22,000 to 23,000 people. We have begun to turn to supplying guidelines and technical assistance to these areas. We have asked them to begin to create plans for their areas. Every neighborhood within an NPU area sends a representative to a Neighborhood Planning Council, and that Neighborhood Planning Council has set out to start creating plans for that area.

We have published two guides for citizen use. One is called *The Values of Neighborhoods* and the other is called *How to Do Neighborhood Planning*. We held hearings to determine the needs in each of the areas, and we published what we learned from those hearings, distributing it around the city in a book called *Interaction*. By the end of January, 1977, after a year of effort, the first of our 24 NPU plans will have come off the press. For the most part, these will become input to the five-year plan in the 1978 Annual Comprehensive Plan, which is already being put together.

I would like to conclude by making three observations. First, if the plan for a city is to be valuable and used, it must have at least the two important characteristics of authority and implementation. It must have authority first: it must compete on an equal footing with the budget. That means it must be current, it must be relevant, and it must be adopted policy. These are the essentials that make the budget authoritative. The budget is debated each year, often furiously, each step along the way. Then it is adopted by law. When the budget changes as required, it does so as a result of amendments to the law, and again these are argued out, negotiated, compromised, and finally accepted or rejected by vote of the Council. The comprehensive plan, if it is to have influence, must move along these same lines and have these same qualities. It must be annually or biannually prepared so that it will be relevant. It needs to be widely debated, so that we have our priorities agreed upon, and then formally accepted by our governing body, so it will have legal authority. Plan-making time is when the priorities ought to be set, not at budget-making time as is often the case. The reason why the plan usually has such little authority is because the priorities are set when the budget is set. The priorities ought to be set when the plan is set, and that plan can only be set if it is relevant and if it is legal.

The plan must be prepared so it relates to the needs of those who have to implement its recommendations. For example, Atlanta's first year plan was a typical plan. It set forth the goals and objectives and the plans for achieving these, but it was a product that principally served a planning rather than a management function. Hence, it had little effect on the department's activities because so much of the language that planners speak is foreign to bureaucrats. To make the plan understandable to the departments, the current (1977) plan is broken down into categories that are not the traditional planning categories; instead, they are structured in categories that are consistent with the structure of the executive branch. Recommendations in this year's plan are expressed in a format that permits them to be numbered and then tracked as they move into the department budgets, starting with the goals and objectives, actions, and programs of the city in the departments that are designed to carry them out. In other words the plan and proposed program budget, which we are also working on, have a common format, so that one can flow directly into another. The bureaucrat down at the far end, who is carrying out the projects, can follow back through to the comprehensive plan and understand why it is that he is doing what he is doing. This allows departmental actions to be related to the larger purposes of the city. So to summarize the first observation, we need to make the plan useful, and in order to do that it must be authoritative and consistent with the budget formats.

The second observation I want to make has to do with citizen participation. Whether we like it or not, citizens in many places want their voices to be heard, and today they are smart enough to have adequate time and resources, and are backed by sufficient legal powers, to effectuate their desire for participation. In Atlanta I believe that their arrival on the scene and their voices in the deliberation have enormously improved the quality and relevance of the decisions we are making. It would be derelict not to recognize and clearly understand the limitations of citizen participation, however. For the most part citizens are not organized for the long pull. They tend to react without sufficient information, and their opinions tend to be highly parochial and personal. They often are together when threatened, and when the threat is removed they wander away. In the final analysis, it is the elected official, established by the system, who recognizes and represents the concerned citizens. Rarely should citizens' voices be a substitute for that of the elected official who is ultimately responsible for the welfare of the citizens and city.

The final observation I would like to make is that in order to make city plans effective we require new and different kinds of planners than those we have grown accustomed to. Most urban planners are taught how to produce plans, and little emphasis is placed on how to get those plans implemented. The low value that many officials place on the work of the planner often grows out of the officials' perceptions of the meaninglessness and uselessness of the plans that we as planners draw up. We should know the plan is only the first stage—rather than the end of the process—of our duties in producing a program that is implementable. The test of good plans is largely but not totally whether they are followed. The future planner should know the techniques by which a plan can be accomplished, and I would urge every planner to concentrate on plan implementation.

QUESTION: How would you characterize planning in Atlanta by expressing to us the single most important objective of the planning function or that goal which best characterizes the planning function?

RESPONSE: I have two. *The first one is the idealized function, that is, we are trying to make the city tolerable to live in, and, I guess, the basic common denominator there is to design a city where you can raise a child.* To make a city habitable and enjoyable, tolerable if necessary, one must be able to perform the functions that the people demand of it. *The other function I believe is the management function; that is, to help guide the decisions of the city, of the various departments of the Council, and so forth, in such a way as to direct them toward a common goal,* which is the goal I believe the planning function is trying to set—a goal to produce the kind of city I just described. It is planning that can begin to direct the decisions so that the resources will be productive, consistent, and directed toward the goals that the city sets for itself.

Q: We know that Atlanta is currently building a heavy rail subway system that is somewhat regional in nature and similar to the Bay Area Rapid Transit District in San Francisco. Melvin Webber has characterized the BARTD system in San Francisco as one in which the rich ride and the poor pay. Atlanta is relying to some extent on the sales tax to pay for the construction of the system. What are the planning mechanisms in Atlanta, either within the city or on a regional basis, that exist for seeing that the same kind of characterization cannot be made of the Atlanta system?

R: In the first place, the BARTD system and the Atlanta system, even though they were designed by the same engineering firm, are considerably different (partly because of the way they are financed). BARTD had no financial mechanism that would insure the system being built on a timely basis. They had to go hat in hand for every dollar they got. Atlanta does not have that. We have the one cent sales tax. It generates something like $4 million a month. We have committed ourselves to a single fare. BARTD has an enormously high fare: it is for the rich! It costs $2 or more to go from downtown San Francisco to Fremont. It costs 15 cents wherever you go in Atlanta. If you want to transfer, the transfer is free. It is designed to serve the lower income people, and we are able to do that because of the committed and continuous tax. It is almost an inflation proof tax. When the price of things goes up, the money still comes in.

Melvin Webber might say, "Well you've used a regressive tax," but, in fact, Atlanta is the third largest convention center in America and between 30% and 40% of our sales tax is paid by out-of-towners. So it is an exportable tax, and it has been a very good tax.

As far as the other contrasts with the BARTD system, and with every other system I know in the world, we have had more urban planning attached to the design of this system than any other system. It began before the system was actually set in concrete, and it greatly influenced the location of stations and what we were trying to achieve. One of the things was service to the poor, and another was trying to put stations in urban renewal areas that would generate new activities. That began back in 1967, and there has been continuous planning with the help of Urban Mass Transit Administration (UMTA), which gave the region $500,000 each year for three years. We created our station area plans before the system was finally designed, and those plans changed the designs of the system. The transportation planning and general planning went along at the same time, and one was able to affect the other.

In our planning program, 27 of the 41 stations in the system are within a single jurisdiction—the City of Atlanta. We created citizens' advisory committees around each of these stations, and they participated in the design and station area planning. We have created what we call an urban framework policy, which is a growth strategy for concentrating development around certain stations, and citizens have participated in that.

So I think we learned a lot from San Francisco. By the way, I have personally visited there at least once every other year since 1967 when I began working with the Atlanta system. I think we learned a lot, and many of the problems that BARTD has experienced will not be repeated in Atlanta.

Q: You were talking about trying to make neighborhoods understandable, like an area that a child could comprehend. Then you said when you did this you came up with 190 of these areas and you could not comprehend those, so you came up with 24 of them. Is not that sort of a contradiction?

R: We have not gotten rid of the neighborhoods, and the neighborhoods continue to plan for themselves. What we found is that when we tried to work with all these neighborhoods, there were just too many for us, and they tended to fight with one another. We still encourage neighborhoods to create their own neighborhood plans and to design this identity concept that I tried to put across. At the same time these neighborhoods are working with each other to produce area-wide plans: the NPU plans. If all these neighborhoods want a park, for instance, somebody has to give priority to the park in that area that seems to be the most important. We like for the people out there to fight it out among themselves rather than to have us decide it. That way they tend to pit one another against one another, but the neighborhoods know where the parks are needed, or where a stop sign is needed, or where rezoning ought to be started. So they are working to create a plan for their whole area, but individual neighborhoods are still creating their own plans. We got out of the neighborhood planning business, but we are helping them do their own planning. We are concentrating on big chunks of the city where we think we can be much more effective.

Q: What sort of interaction do you have with the regional and state planning agencies that you come in contact with?

R: Far too little! The country's oldest, continuous, regional planning agency is in Atlanta (Atlanta Regional Council). It began in the early 1940s. They create regionwide plans, and we participate in the creation of their plans. In fact, they pick up our planning and put it into their plans as much as they can. However, they tend to deal with issues that are on a regional scale, that only sometimes apply to our planning process. We tend to be working at a different order of magnitude in our area. Ours is a much more direct contact with people. It is much more concerned with zoning, land use, and tangible things. They are talking about systems. We participate in that planning, but I do not think there is much conflict. Sometimes we conflict on a single road project, but by and large they have gone their own way with dealing with the larger programs. We tend to deal with a different order of problems.

We have very little contact with the state. If there is a hospital, highway, or a facility of that type, we get involved with the state. We are doing a joint study on a ride-sharing program in what we call the Capitol Hill Area of the city (the city, county and state occupy the same space in Atlanta). The Georgia Community Development Department works out in the rural areas, however, and they have very little to do with the City of Atlanta. The city and state tend to go their own separate ways.

It is unfortunate that this is the case both with the region and state. We are one-third of the region and a big chunk of the state, but they are simply not prepared to operate on our kinds of problems. We are unique. We are by far the largest city, and they are not geared up to participate in areas where we need help.

Q: One of the big criticisms of planning in recent years has been that the planner has been isolated from the community. With this criticism in mind, and with the citizen participation in Atlanta in mind, who is responsible for soliciting this citizen participation—the planners or the politicians?

R: My feeling is it ought to be the politicians. My own experience has been that a number of the politicians—not all but a lot of the politicians—are very careful about their

[handwritten margin note: So how was this politically made feasible?]

dealings with the citizens. For whatever reason, they do not want to stir up the rabble, or they do not want competition, or they just do not believe in the process. It is that simple. We have to go behind them to get the citizens to organize themselves. The Citizen Participation Ordinance laid all this out on how the NPUs will be organized and their functions. The fact is that citizens write their own by-laws and do their own thing in Atlanta. The Ordinance said the Council member from the area would be the person who would organize them. The Planning Bureau was not to organize them. If they were organized, we could go in; in fact, we have had to go in and help pull together this whole thing in a couple of situations where the Council member simply was not interested. We avoid trying to do that.

My planners—I have told them on a number of occasions and I continue to tell them—are not advocate planners. I have told them that they are making a big mistake when they try to be advocate planners because that means they are working for two different people. When you work for two different people, you are in trouble. They work for the city, and they give technical assistance, advice, help, and bring back the concerns of the citizens. They are professionals first and foremost, and they are to represent that point of view objectively. They have to give me their judgment of what they would recommend for that area—not as a neighborhood advocate, but as a professional planner. It is up to me then to pass that judgment on to the Mayor, and it is up to him to pass it to the Council. That is very hard for the planner who gets emotionally involved. That is the problem of working in the neighborhoods. The planners tend to get emotionally involved in their work.

Q: You mentioned earlier that the voices of the citizens should not be a substitute for those of the politician. Are the voices of the citizens in Atlanta hurting your comprehensive planning efforts in any way?

R: I think that the point I made was that ultimately the citizen representative is the elected official, and it is a mistake to put too much reliance on the citizens and what they have to say because the citizens tend to be a very amorphous body. They disappear, they come back, they change, and so forth. That is why we have elections. We can structure the system so that a person serves, so we know who is the speaker because he or she is the person elected. We know how long these people serve, and we know how to replace them; there are rules of the game.

The structure is the ultimate thing that we have to depend on when we make decisions in the city. The citizen's voice is extremely important. I have tried to make the point that I consider that an important element. But there is no substitute for that official voice, and if we get carried away with our egalitarianism and believe that all this citizen stuff is the end-all, we are going to make a big mistake. We would be leading ourselves down a primrose path, and we would be very upset about it in the long run. It is one factor. There are special interest groups, and there are concerns other than what the people living on these streets have. It is the planner who needs to wade through and then make the judgment. All I want planners to do is to make certain they listen to the citizens very carefully and give as much weight as they can to it. But ultimately they have to weigh a lot of the considerations and come up with a professional choice.

Q: How are you enticing people back into the inner-city neighborhoods?

R: I would like to take credit for all those people coming back and there are a lot. Thousands of families are coming back. Actually they are not coming back; that is a very bad expression. They were not there before! These are new families that are moving into the city. They have a choice of moving to the city or suburbs, and they have opted to come to the city. Most of them are not coming back. The "empty nest" people, older families where the children have gone, are moving back, but by and large these are young people who have just gotten married and are settling in.

I have to say that as much as I would like to take credit for that, and I will probably take as much credit as I possibly can squeeze out, the major factor for bringing the people back is the city itself. The excitement, variety, and nature of the city are attractive, and people like this in contrast to the suburbs, which they may not like. There are also the economics of it. The housing is cheaper in the city. There are big houses that are empty or that are cheaper; they can buy them with their own equity, build upon them, and put them in working order. I think that is a major factor. There are also the economics of travel. They have to live further and further out, and the city is beginning to be able to compete for the first time with suburbs. (The FHA notwithstanding—we can overcome even that.) The economic trends have begun to slip back toward favoring the city and are giving us a little competitive edge, and I think that is bringing them back as much as anything.

Once they are here, I think we can be helpful. Central Atlanta Progress, which is the downtown private planning agency, has gotten together the banks and savings and loan associations and created a pot of mortgage money. This is slightly higher risk money, about $62 million, which is a tremendous sum to put into the neighborhoods to make lending a lot easier. I think that is bringing them back, too. But other than that, once they are back our role begins, and that is where we can begin to clean up that area, to police it, to put sidewalks in, to plant trees, and to do all those things. Our role is not to do the huge clearance we did in the 1960s. I think that is an era long gone, one that will not come back. We overgrazed the land, and we will not do that again.

Q: You talked about the importance of the new strong executive. How much of a role does the Council actually have in formulating the comprehensive plan?

R: They hold their own public hearings. The plans begin to flow from our office and from the neighborhoods, and we put together these plans. We then go back to the neighborhoods and hold hearings all over town. We say, "Here are the plans that we are going to present to the Mayor. We will give you one more crack at it to see if you object to them or approve of them." Then they come in, and they look at them, and they make certain suggestions. We accept these or we change them. Then we send the plans to the Mayor, and the Mayor sends them to the Council, and then the Council has its own public hearings. They go through the same process. The council member generally comes to the public hearings in the neighborhoods, and then he or she comes when the Council holds public hearings. So they usually are directly involved at two stages in the process.

Q: What changes have occurred in Atlanta regarding the racial structure of the city and the distribution of families of various races as far as residential and business uses are concerned?

R: There has been quite a bit of change in the last fifteen years. The city is approximately 60% black. We have a black Mayor, and of the nine commissioners four are black. We have a black superintendent of schools, and two Congressmen are black. About 90% of the student body of the schools is black. The blacks have moved into formerly white areas, but the housing patterns are still segregated. I think this is so more by choice than anything else. Ten years ago blacks moving into a white area created panic, but now I do not think that would raise many eyebrows. I do not think there is much discrimination in housing. There is a case here and there. For example, a white realtor may not show a black a house in a white area. But I do not think that the amount of discrimination is high, and I do not think that is the major factor for maintaining the separate racial pattern. I think it is voluntary and partly economic, because the white areas that remain white tend to be fairly well-to-do and the number of blacks who are in that economic class is not high. There are some extremely fine middle and upper class black areas. Atlanta has probably one of the wealthiest black communities in America, so there is not a lot of push toward housing integration.

Q: Most of your presentation dealt with systems, and I find planners by and large are systems oriented. In practice, however, personnel sometimes spells success or failure. I wonder if you might comment on some of the problems or changes that have occurred under the new Charter for personnel and the bureaucracy.

R: There was and remains an enormous problem of converting to a strong executive system within the bureaucracy. In fact, I would say that is a bigger problem in the bureaucracy than it is with the Council. The Council was suspicious of losing power, but it was the bureaucracy that had remained independent for all those years. They suddenly began to feel that they had to report not to their Council member, with whom they had become familiar, but to the Mayor, whom they did not even know. The Mayor reviews their programs now. It is going to take years for this to change. They went through a budget hearing this year for the first time before the Mayor. They had never had a hearing like that before, and they had to defend their own budget. They had never done that; the Director of Finance had done that. It is a very unfamiliar role.

One thing that made it a little easier is that when we consolidated 26 departments into nine, and everybody was reporting to a commissioner, the former department heads became bureau chiefs. They were now reporting to an unfamiliar commissioner, but it was within their own department. We could not have made a more drastic change in our government. It was a drastic thing, and a bureaucracy does not respond well to reform. But it can be done, and I am sure we will make the changes work.

Q: Does your department have an overriding philosophy on how to deal with its constituency?

R: I think that planners have to be involved in implementation. They have to understand that when the plan is produced, it is only the first step. They then have to figure out how they are going to carry it out. They have to write ordinances; they have to lobby; they have to be willing to change; and they have to refine, interpret, and cost-out.

Planning is a political process. The person who does not realize that every planning decision is a political decision is not going to get very far. We are in a political environment. We have always fought to be next to the executive to influence policy. To be a policy advisor means one is in politics and cannot avoid what is going on politically. It is a very difficult and turbulent kind of life. If we are really going to get our plans accomplished, however, we must get involved in politics. We have to do our homework and fight for our plans, but we also must take those plans and see that they are carried out.

The constituency, the people that the planners work for, are the people who pay the salaries. The constituency is the city government, and they cannot be the employees of the citizen groups. They can get themselves in trouble by being advocates because they cannot deliver what they are advocating. I am all for neighborhoods having advocates. I would be happy if we could make money available so neighborhood groups could hire advocates. That advocate cannot be paid by one source that is trying to articulate the concerns of all the people, however. We are part of the establishment when we are in the city government. It is to have sold out to the establishment. We have to decide where we are and where we cannot play. We can get ourselves and the city in a lot of trouble trying to do both. I have seen it happen a lot of times, and the planners end up hurting the people they are trying to help.

Q: Are you having any conflicts between existing residents in the neighborhoods and the new type of person moving into the city?

R: There is a large number of such conflicts. In most of the eleven reviving neighborhoods where young people have moved in, they immediately begin to agitate and become activists. In almost every case, they suddenly come into conflict with the existing residents of the neighborhood. They have problems within the neighborhoods on who speaks in behalf of the neighborhood and on what they want to emphasize. Usually however, the young activists win.

Q: New York City's new Charter also mandates neighborhood planning. It provides for a paid staff for that neighborhood planning, with the staff being controlled by the neighborhood, so there is no need to worry about city planners being neighborhood advocates. There are separate neighborhood planners whose job is to be neighborhood advocates. Do you see Atlanta moving in that direction?

R: I would be happy if Atlanta could afford to move in that direction. New York may be able to afford it, but Atlanta cannot.

QUESTIONER'S RESPONSE: We hoped it could!

Chapter 3

CLEVELAND: PROBLEMS OF DECLINING CITIES

NORMAN KRUMHOLZ

Executive Director
City Planning Commission
Cleveland, Ohio

I would like to talk in general about five different points: (1) the general background for planning, at least from the perspective of a planning director; (2) the Cleveland approach and how it may differ from some of the other approaches that have been used in other planning agencies; (3) some innovations that we may have adopted in practice; (4) some success stories, and, unfortunately more frequently, some stories of failure; and (5) some lessons learned that other professionals in other planning agencies might be able to use.

BACKGROUND

Let us start with the general background for planning, and the very critical perspective of the planning director. Assume that you are a newcomer in a city and your assignment is to be planning director of that city. Normally you would want to set your mark on the work program of that agency and do some things that are of interest to you and that you consider to be particularly useful. How do you construct a work program? To begin, you have to continue what is being done that is mandated by the Charter of the city or the enabling legislation that set up your organization. In Cleveland, an example of this is zoning, which does not take a great deal of time but must be continued. Other examples include mandatory referrals and some capital improvement programming.

In addition to those things that must be done by Charter, you have to continue what the staff or part of the staff has been accustomed to doing or what powerful constituencies in the city expect you to continue doing. An example of this in Cleveland is design review. We have a Fine Arts Advisory Committee, which in my view takes up a good deal of time and does not really produce anything of tremendous importance. It is expected to do this, however, as a function of the planning agency. The newspapers particularly put great stock in what the Fine Arts Advisory Committee has to say about trees in front of public buildings, façades, and that sort of thing. So we continue with the design review activities. But the director may also, if he is aggressive, define new purposes and new objectives for the agency.

How does the director come by this freedom? After all, police agencies cannot redefine their function or purpose; people who pick up trash cannot do that; people who run the parks and recreational divisions cannot do that. But I would argue that the planning agency and the planning director can redefine their functions and their purposes. The director comes by this freedom in several ways. First, many of the traditional tasks that are underway occupy very little time. The kinds of things that we have to do in Cleveland would probably occupy ten to twenty hours per week, and the director would then be through with those kinds of activities.

Next is a fact of life that most people experience in the work world. The structure of city bureaucracies—and, I suspect, the federal and state government bureaucracies—is not monolithic. It may appear from the outside to be that way, but there is, in fact, a very loose structure, and there are a lot of unclear or undesignated responsibilities. For example, in the City of Cleveland there is no single agency responsible for developing policy for highways, and there is no single agency with a delegated responsibility for regional planning coordination. No *we* do all these things! The reason we do them is because we choose to do them. We have demonstrated enough competence in all of these areas to be considered as serious players in the game.

Next, planning practice is not uniform at all by law or custom. Those who have worked in planning agencies know that the work programs of individual agencies can wander all over the place. When I arrived in Pittsburgh, Calvin Hamilton, who was removed shortly thereafter, was running the agency. At that time Pittsburgh was doing the Pittsburgh Simulation Model. Cal and his boys were the darlings of the academicians. They had this mathematical thing that was going to forecast the state of the future and all sectors by the year 2020. When Cal was forced out of his job, the new director was straight out of the Democratic Administration and the Chamber of Commerce. He was John Mauro, who was not a trained planner but rather a business entrepreneur-hustler. The face of the agency and the importance of the agency changed radically overnight. From an agency that was considered to be the darling of the intellectual community, when Hamilton was running it, and that had no impact at all on local decision-making, because nobody on the Pittsburgh City Council cared what Hamilton and his people were up to, it became entirely the reverse. It became under Mauro's direction a very important agency to the City Council and Democratic Administration and not important at all to the academics on the cutting edge of the profession. So the individual, the person, in every case, will shape the work program of the planning agency, sometimes in radically different ways.

I guess the final reason that a planner can redefine operations in local government is that, in most communities, most of the people in local government have no idea at all what planners do. There is a general notion that planners occupy an ivory tower somewhere. There is a notion, often reinforced by some of the activities that planners engage in, that they are the forlorn defenders of backwater issues that everybody else in the town has forgotten about or certainly does not care about any more. So given the fact that nobody knows what a planner is supposed to do, one may say, "Hey I'm doing this because I think I should." Everybody will take his word for it. It is as easy as that.

The selection of a work program, and the selection of activities that go into the work program, it seems to me, should depend on a number of items. First, the ethics of the individual planner and to a lesser extent of the profession. Second, the physical, economic, and political realities of the locale. Change in the operations of city planners, as in the operations of all public bureaucrats, comes from increases in the interest and the political power of consumers, not from the good intentions or the moral virtues of planners. Third, the thing that shapes what we all do in the field is client demands. Planners without a strong internal drive to do certain things will mostly do what the people and their politicians insist they do, if they in fact insist. If nothing is demanded, the planners may become both the producers and consumers of their own products, and that is a problem.

One of the things we do in Cleveland is seek out latent consumers and latent allies for our own activities, and then we energize those people to demand more of us. There is a great symbiotic relationship that can be developed in this way. Our work program in Cleveland depends on these three imperatives: (1) ethical and moral considerations, (2) client demands, and (3) the very powerful realities of the situation.

What are the realities in Cleveland? Before I go into the bad news side, let me emphasize that it is distorting and not fair to oversimplify the realities of the city. There are many strong aspects of the City of Cleveland. There are many good neighborhoods that I could show you. There are about 15 million square feet of rented office space in the downtown area and the occupancy rate is very strong. Cleveland ranks behind only New York and Chicago in headquarters for Fortune 500 corporations. The City of Cleveland still has about 40% of all the jobs in the Cleveland metropolitan area. So there are very strong, powerful economic forces still operating in the city. But the long-term trends in Cleveland and similar cities spell out municpal decline and personal poverty to a great extent.

The city's total population is falling very rapidly. It dropped by 39,000 in the 1950s, and in the 1960s it lost 125,000. It is now falling at a rate that we estimate at 121,000 per decade. The percentage of blacks in the city's population went from 16% in 1950 to about 40% today. The white population has fallen about 50% since 1950. A rising share of the city's population is economically dependent, and these people's income is falling in relative terms. About 20% of the city's families now receive Aid to Families with Dependent Children, and one-sixth of all the families in the City of Cleveland have incomes under $2,000 per year. The more affluent families are departing, which is not new, and in the 1960s the city lost 25% of the families with incomes over the median for the Cleveland metropolitan area.

The city has a declining assessed value base. That is to say, our ability to meet the needs of this increasingly poor population becomes more and more limited. Our assessed value base fell 5% from 1969 to 1974, while at the same time the consumer price index rose 34.5%. The local general fund operating revenue has declined 2% in current dollars for a 37% decline in real terms over that same period. It is only the rising federal transfer payments that have maintained the city's purchasing power. The employment base for the entire Cleveland metropolitan area is growing more slowly than that in the nation as a whole. The city lost 71,000 jobs, or about 15% of our whole job base, during the 1960s. Our economic base consists of slow-growth industries. According to 1970-1973 census data, the whole Cleveland metropolitan area is declining slowly in population by about 1% a year. This is different from central-city decline that occurs while the metropolitan area grows. So as long as there is growth in the metropolitan area, presumably there could be a tax on that additional growth if the jurisdictional reach could go out there. If the whole metropolitan area is declining, however, it is increasingly doubtful as to whether the central city can grow back to health.

Cleveland has a high and rising crime rate that creates insecurity—which in turn discourages people from staying here or moving in. In a survey we conducted recently, crime was seen as the number one local problem by 73% of all our residents. In 1974, the rate of violent crimes against persons per 100,000 residents was 1,730. That was 16% above 1970 and 164% above 1965. Yet as we know from looking at the Census Victimization Study, the actual crime rate was 4,150 per 100,000—2.4 times the reported rate. We have some severe doubts about how effective the service package of our safety forces may be. There is this question even though the Cleveland safety forces are far better paid than those of surrounding municipalities. The question must be asked: How good is the service package that we are buying for that money?

Another problem area is rental housing. Because of rising costs of operation and ownership to private landlords and a declining ability to pay among lower-income tenants, there is an economic squeeze on rental property owners. Most of these rental property owners are not large landlords. They are not the prototypical rich, fat suburbanites grinding the faces of the poor. These landlords are mostly small owners and mostly owner-occupants. These people are in a bind, coping with the squeeze that arises from having to deal with a market that has increasingly low incomes. They are resorting to tax-delinquency first and to absolute abandonment next. About 8% of the properties in the city (about 12,000 parcels) are now tax abandoned. Most are in the poorest quality areas. Abandonment by renters and owners alike is well underway in the West Central and Hough neighborhoods.

How do planners and planning agencies typically respond to the constant imperatives I have laid out? I would argue— and this is based on a very limited sample, plus some reading in the field—that they respond by doing business as usual—in effect, by ignoring a lot of these problems. They respond by executing visionary futuristic plans; by doing land use schemes for the year 2020; by arranging design schemes in areas where there is no demand for land; and, where they have no control over either the public or the private side of the development, by confirming the need for bike paths or whatever seems to be interesting (and that the federal government is funding at the moment). It seems to me that most planners tend to assume that their activities—that is, their professional planning activities—are beneficial to the public. I think we have to ask whether that is really the case. It seems to me that many plans that I have seen (and that you have seen) are actually completely unrelated to actual development. They are repetitious in style and content. I am sure you will note in your studies the similarities of most of the general plans of most cities which all call for growth—economic growth, population growth, or, at the very worst, stabilization. If there were some unique aspect about planning a realistic future would not some of the plans of the 1960s have forecast, instead of the growth of the 1970s, what actually happened—i.e., the decline of the 1970s? It seems to me, as Jerry Kaufman has pointed out in an article on American planning practice (pp. 111-138), that there is a wide discrepancy between what planners are being taught in graduate schools and what planners are actually practicing in the field.

THE CLEVELAND APPROACH

Let me move then from this general background of planning to the Cleveland rationale. The advice, recommendations, and information offered to policy makers by the Cleveland City Planning Commission and its staff are simply aimed at the accomplishment of one single goal, and that goal is very easily stated. *Equity requires that government institutions give priority attention to the goal of promoting a wider range of choice for those Cleveland residents who have few if any choices.* This is a very narrow kind of objective. We feel the goal arises very naturally, that is, it is not really ideological, although it may strike some people as being ideological, but rather it arises from the reality of conditions in the City of Cleveland. It arises as well from what we take to be egalitarian ideals that are inherent in our country's religious and political beliefs. It seems that in a city like Cleveland, which is characterized by relatively low incomes and the problems that flow from poverty, working to improve choices for those who have few may in fact be simply providing appropriate services for a large and growing sector of the population. This is a sector that forms a political constituency (which may also be a reason for my own survival through two very different administrations).

The goal that I set up also provides a focus for our day-to-day efforts. It enables us, the staff, to identify issues that merit priority attention. It provides a framework for staff analysis. It allows us to focus on the essentiality of an issue. We rarely have problems in defining the core of an issue. Our goal is essentially redistributive, and we ask of the programs that come before us (and the programs we develop ourselves): who gets and who pays? When proposals appear to offer more benefits than costs to city residents, particularly the poor, we endorse them. Those that do not are questioned and often publicly opposed and lobbied against. So it helps us to focus on what we perceive to be the essentiality of issues.

There is very little in terms of an essential theory of planning that went into the formulation of this goal. As a model, I suppose, it is close to Paul Davidoff's advocacy framework (pp. 331-337). But we take our program beyond the goal's statements and into politics and action. That is, the active, continuing effort to shape the decisions of politicians, bureaucrats, and public and private entrepreneurs toward our direction is more than merely making advocacy statements in the pages of the *AIP Journal.* Day-to-day efforts to really bend decisions, real decisions and real futures, in a direction we want is the most important dimension of the Cleveland operation.

SOME INNOVATIONS

Let me turn briefly to innovations such as they are in practice. I do not know that we have many innovations to report. We have adopted a priority goal that focuses on equity. I guess that is an innovation, but advocacy planning in such a procedure is no stranger to any planner. We adopted the goal—for those who have been paying attention—by ourselves. We did not set off on a Goals for Cleveland program. We did not go into a heavy citizen participation effort to find out what the citizens wanted. We sat around for a couple months and said, "Hey you know, really this is what we ought to do." That is what we are doing now. So maybe it is an innovation in an era characterized by attempts toward citizen participation.

I guess the interventionist approach may be an innovation, because there is some danger down that road. The use of relatively high-quality data-gathering and analysis is not an innovation in planning. I suspect that it may be that planners are in love with technique, but we rarely are willing to go beyond technique into the decision-making arena. I suspect that the staff analysis is of high quality and innovative in that sense. I am confident that what we do in Cleveland is better than anything else in town. That is how we have an entrée into the work that other agencies are attempting to do, those programs we are attempting to shape in our direction. I guess our attempt to focus on decision-making with relevant information over a long pull in a clear direction is perhaps an innovation.

SUCCESSES AND FAILURES

Let me briefly describe some of our successes, and, also, some of our failures. I think one of our successes had to do with a highway proposal, the Clark Freeway. It was authorized by the regional planning agency over our vociferous objection. We produced an analysis that showed the proposal offered practically nothing to the City of Cleveland at considerable costs to the city and the people of the city. Our recommendation to then Mayor Carl Stokes was that we challenge the regional agency and try to get them to rescind that highway decision. This led to a suit in Federal Court and administrative complaints to the decertification of the Northeast Ohio Areawide Coordinating Agency. This is the only decertification ever of a regional planning agency by the U.S. Department of Housing and Urban Development. Ordinarily audiences of planners usually break into cheers and wild laughter at this report. Subsequently, after Mayor Stokes left, Mayor Ralph Perk made up with the regional planning agency, but not before we put eight more representatives on the agency board (which I have sat on for seven years now) and not before they rescinded the controversial Clark Freeway proposals. So I think we have to consider that kind of a major win.

In addition, in the spring of 1975, the City Planning Commission Staff and I created the framework for the negotiations for the transfer of the Cleveland Transit System to the regional transit authority. The politicians in City Hall originally wanted to settle merely for control of the regional authority board. This is a kind of simplistic notion. It is a fiction, of course, because they assumed that if they had control of the board they would control the agency. Baloney! Board members are appointed for six years and, as soon as they are appointed, they forget who appointed them and go about feathering their own nests. We changed that very simplistic notion (of six appointments for the city on a ten-member board) to one where we were successfully able to bargain for reduced fares. We have a 25-cent fare now and reduced fares for the aged and the handicapped who are heavily transit dependent. For the first five years of the operation, those people ride free for twenty hours a day, and they ride at half fare during off-peak hours, four hours a day. We got increased service in terms of reduced headways, that is, vehicles run more frequently. We got tighter route spacing, so there is less distance to walk for a transit vehicle. We got a supplementary service called *Community Responsive Transit*, which resembles dial-a-bus right now, in the transit dependent neighborhoods of the city. We think we worked out a good deal on the transit negotiations, and we were very heavily involved in that for a long while (not without some pain I might say). But it was a long battle, and it is one that we successfully brought to a conclusion.

In addition, my staff has worked for the last four years on a project that culminated in a change in state legislation on the whole problem of abandonment—that is, how to pick up the title on abandoned parcels, clear the title, assemble the parcels, and merchandise them some day when the land becomes ripe again. We quantified the dimensions of the problem, got foundation funds, hired a technical consultant, and drafted the change of the state legislation in House Bill 1327. We then lobbied the bill through the state legislature, which passed it on the last day of the general assembly. That is now law, and that, more than anything else, is completely ours from beginning to end. We even got a book out of the experience.

Those have been a number of important successes, but we do not always fight and win; we get bombed a lot more than we win, and we lose maybe 80% of the deals we get involved with. Let me recount some of the ones we have lost. The first issue was a rezoning issue. When I first got to Cleveland, a developer wanted to impose very high densities for residential development on an abandoned amusement park that faced Lake Erie. We insisted, as a condition for the rezoning, that he dedicate 300 feet of the depth of the frontage on the lake to the public for a park. Our insistence, of course, was illegal, but nobody cared about that. Anyhow, we successfully got the City Planning Commission to disapprove the zoning proposal, but we were overridden in the City Council by a 32 to

1 vote. The one member who supported us did so for the wrong reason; he had thought a different bill was coming up, and he voted the other way. Then we got involved in disapproving a proposed $350 million downtown tower investment because we felt it was a phony deal; the city was going to be asked to make massive capital improvements that were hidden in the legislation with no *quid pro quo*s. That is to say, there did not seem to be any new jobs for city residents, particularly the unemployed. There did not seem to be any new tax ratables, since the developer was going to go out on a tax abatement scheme, and we were going to be held for $15 million in on-site public improvements that we did not have any obligation to provide. We opposed it, and the City Planning Commission disapproved it. The President of the City Council called the City Planning Commission a "bunch of baboons," called for my resignation, and so on. Then the City Council overrode us—I think by 31 to 2. We have 33 city councilmen, so we did a little better on that one.

We also indicated some years ago that the police were assigning their manpower in a way that was not relevant to the incidence, either temporal or spatial, of crime. Officers were assigned evenly around the city, and the crime was largely taking place on the East Side. So we made some strong recommendations and almost literally got shot full of holes by the police, who have to be the toughest outfit in town (with the possible exception of the Board of Education).

I guess another major failure is one that looked very promising early in the year. We proposed that the city's small Municipal Light Plant condemn the Cleveland Electric Illuminating Light Plant (investor-owned) and take it over for public purposes. This was a terrific idea, and I still think we should have done it. But we were not able to bring it off, and I am afraid that poor Muni-light is going to be run into the ground through city mismanagement and a fifty-year campaign by Cleveland Electric Illuminating to drive out the public competition. Muni-light is now nearly dead, and I think the administration is currently making a deal to sell it, which may or may not be accepted by the Federal Power Commission. Up to very recently, we had a case in Federal Court charging anti-trust practices against Cleveland Electric Illuminating; it looked quite promising for a while but now looks like a failure.

LESSONS

What have we learned from all this? City planners and other public administrators—and I think a city planner is one form of public administrator—frequently tend to view urban problems simply as a source of employment and institution-building. I do not have any objection to that, frankly; a lot of us do that sort of thing (and I think it is too bad, but so it goes). I think we have done pretty well as a profession without being particularly useful to the people who need our help. If we were really serious, however, we would be making a contribution by making things better for the people of the cities, particularly for those people who need our help the most.

It seems to me that we should have understood a long time ago that this challenge requires much more than our typical response. The planner and public administrator in general want to address a problem only in terms of their own professional skills, and then they want to stop. They do not want to go beyond what they perceive to be their own professional skills. If a person is a designer, he develops his urban designs or builds a simulation model, and it runs or it does not run. He identifies apparent solutions to the problem, takes those apparent solutions to the client—which in our field is usually the City Planning Commission—presents them as lucidly as possible, and then stops. He stops because his stomach tells him that what lies beyond is dangerous. In other words, the planner in that kind of situation develops a proposal but is not willing to see that it is adopted or implemented. He sort of throws it up in the air and leaves it with the Planning Commission—made up of seven members who have their own businesses to worry about. He leaves the implementation, adoption, and guts of the thing to others who will lack his knowledge, lack his interest, and probably lack his zeal (if he has any zeal). That is clearly not enough.

It is not enough if we are really serious about affecting outcomes, and that must be part of the definition of a city planner. Is not a city planner one who is interested in outcomes and ends and futures? Sure he is! But planners do not decide important issues. Planning commissioners do not decide important issues. Public administrators do not decide important issues, at least not in matters of public policy. In matters of public policy, the people who decide are politicians, and those who propose ends and say they care about outcomes must come to care about means as well. It seems to me we have to be prepared to spend some time and take some risks in the political arena to work toward the goal of improved city conditions. If we are serious about our work, we have to understand much more clearly that decision-making is a long-term, continuing process, and so is implementation. It is not a single, clear act that just happens in time, after which we may go back to sleep. The Cleveland Electric Illuminating contest mentioned earlier took four or five years to develop. The highway suit took five years to develop, and it is still not fully over. The transit issue—for which I was Mayor Stokes' alternate on the transit coordinating committee, which was set up in 1970 to begin working toward the transfer—finally came to its completion in the spring of 1976. That is not an act; that is a process. It is a process that demands our protracted participation and our willingness to get in the arena and take some risks toward the objectives we say we are concerned with.

The payoffs in our profession, if we are serious about our work, cannot be limited to increases in salaries, or contracts gotten, or tenure achieved, or articles or books published. The payoffs have to be in the actual improvement of the management of cities and in the lives of city residents—

particularly the lives of those city residents who are most in need. So it seems our work is cut out for us. For the future we have to better understand our moral and technical responsibilities as planners within a system driven by powerful ideological and economic forces, a system that does not adapt very well to innovation and that changes only with the greatest difficulty. We must learn to interact with that system and with its political representatives on their terms in their language by talking in terms of issues that will be meaningful to them in their political context. We have to accept our share of responsibilities and risks in long-term, sometimes dreary, sometimes hazardous, day-to-day decision processes. I think if we work seriously and energetically within the context I have described, city planners can play a major role in resolving many of America's urban problems.

This will not come without much conflict. One stimulating thing about the business is that you are always fighting with somebody; somebody is calling you a baboon in the paper, and you are calling somebody else a baboon in the paper, and it is fun in a way. This should not be surprising if one is championing the interests of the poor and the powerless, and one should not be surprised when the rich and the powerful take umbrage on occasion. So there are some losses in the sense that there is always conflict.

There are losses also for an individual in the sense that if you are constantly at odds with the establishment, you may be sure that when the chair at the Chamber of Commerce, or the position of executive director, is available at $60,000 a year with a $10,000 expense account, you will not get it. You also will not get to be the executive director of the Regional Transit Authority at $80,000. You will not get to be executive director of anything that the establishment has anything to say about, and they have a lot to say about a lot of things. So there is some personal loss or a loss of personal options in that way. But you have the consolation of knowing—if it is a consolation—that you are involved in important events. You will be playing the right game, in the right stadium, facing in the right direction, which I seriously doubt most planners are doing today. Those of us who do this may even have a lot more fun. But if we continue to be content with our traditional narrow and risk-free responsibilities, it seems to me we will have very little impact in the future in resolving the real elements of the urban crisis. We may, if we are not very careful, become so out of the habit of providing real kinds of recommendations to real politicians that some day, we may simply disappear with a metallic click.

QUESTION: There seems to be a population in some of our cities that is very difficult to educate. It is very difficult to house that population, and it is very difficult to provide certain services to them. How do you address your programs to provide services to that very population, one that is very much in need but which also might have very detrimental effects on other segments of the city, who might also be in need?

RESPONSE: It very rarely falls out that clearly or that simply. Actually, what we do for many issues is to simply present the countervailing view. We have discussions by staff, for example, on precisely that question. What happens if we were so successful that we just could not possibly believe it? What would happen if all of our recommendations regarding the redistribution of wealth and power and so on actually took place? Would we not be left with a city that would be populated almost exclusively by the poor, and what kind of a city would that be? I am not sure that is a reasonable kind of context. The reason is that we cannot possibly be that successful. Everything in our society goes the other way. To take the position of trying to help those in greatest need is simply to define an opposing point of view. In defining that opposing point of view we can help define their needs in a more prominent way. In cities like Cleveland, Detroit, Boston, Pittsburgh, and Buffalo, and a lot of other cities, there is a political constituency out there that is waiting to be served if planners can only understand in a realistic way that the need is there. If we just try to push a little in that direction, we will get some benefits of a greater or lesser nature for that population, depending on the issue at hand. Whereas if we do not push at all, there will not be anything.

In some of the neighborhoods in Cleveland—the blasted kinds of neighborhoods where everything has collapsed: public services, school system, police force, and everything else—we are trying to encourage the city, through public services, to maintain at least the limit of the city's own standards of health and safety. We are not recommending major capital expenditures in those areas because we feel a lot of emptying-out has to happen before the land becomes ripe again, if it ever becomes ripe. But for the city to extend just its own levels, its own codes of health and safety, would probably strain the ability of the city budget beyond its present capacity right now.

Q: How do you decide what is in the best interests of the population; I am thinking particularly of the controversy that always surrounds the question of pollution versus jobs.

R: We make a lot of intuitive judgments. As I mentioned earlier in my comments, we are not much into citizen participation. Surely somebody is going to slam me on that. But we are really not much into asking opinions of the inhabitants of the neighborhoods of the poor and the working class, for example, which is almost all there is in Cleveland. We are not out to say "Do you want a highway run through your neighborhood?" We do not feel such a thing is necessary for people to tell us that. In the issue of air pollution control versus jobs, if we have any influence on the matter at all, I would say without any question that we would come out on the side of the jobs. That is certainly more of a bedrock kind of issue in Cleveland than are esoteric issues like air pollution control. So there is a lot of intuitive identification of interests of the poor and blue-collar working classes that we make without flailing out and asking them what they feel about a specific issue.

Q: You seem to have little consultation with neighborhood groups. Is there any particular reason for that?

R: We got into that rather early on and got involved with a number of neighborhood groups. At that time anyhow—and this was in the 1970s—they were giving the most antidemocratic things one could possibly imagine as their prime objectives. Frankly, at that point it occurred to me that if we went ahead and followed through on some of the things that these neighborhood groups were saying, things they wanted very badly, we would be in the same position as were the "Good Germans." So we ended that and decided to depend a little more heavily on our own intuitions. I will concede to a certain amount of arrogance in that response, but that is the way it is.

Q: Are you lobbying for new annexation laws?

R: No, I do not think there is a prayer for that in Cleveland. Cleveland, like most older American cities, is surrounded by incorporated municipalities, all armed with powers of land use control that are superior to those of the federal government. All are armed with their own mayors and councils and state representatives and U.S. representatives. There is no way that Cleveland is going to annex Shaker Heights. At least I do not think so.

Q: To follow up on that, Cleveland, from the way you are talking, does not perceive the suburbs as a threat. Could you make a substantial challenge in some way by being able to cut off federal funds?

R: I disagree with that in a sense. The suburbs are a challenge in the sense that the leadership, the civic leadership, that would be in Cleveland, but for the suburbs, is in fact in the suburbs. That is an unquantified loss that is extraordinarily difficult to measure. The reality is that the rich left Cleveland about fifty years ago, and the middle class has largely abandoned Cleveland since the end of World War II. That is not just a numerical loss; it is a loss of intelligence, understanding, and civic and political leadership. It is a very important matter that cannot be quantified. But the suburbs do represent a challenge; our own Sewer Division, for example, was taken over by court order and regionalized as a result of suits to regionalize the system brought in the courts by suburban jurisdictions alleging unfair rate practices. Our Water Division, one of the few divisions that the city still controls that is showing a net profit, is now under similar attack in the courts and very likely will go in the same direction. So there is a continuing challenge. I think in the larger sense, though, that the Cleveland region has to see itself as a competitor. Not city versus suburbs, but rather the Cleveland region versus the Pittsburgh region, for example. I think the realization that growth seems to be coming to an end in the whole Cleveland metropolitan area may begin to stir that kind of thinking in a lot of other people's minds as well.

Q: I am surprised by your comment that you have gotten involved in planning for the police department, transit matters, and that sort of thing. I would suspect you are responsible to the City Planning Commission. Who initiated these studies; Did you take it upon yourself to do that or were you directed to do that?

R: We took it on ourselves. The City Planning Commission in Cleveland has one function—to support the staff!

Q: Do you see any prospects for Cleveland's reconversion economically in order to increase employment in different sectors and increase population and grow again?

R: That is a very hard question—perhaps the ultimate question. I do not know. It depends on many things over which the City of Cleveland has very little control. How expensive housing gets to be in the suburbs, how restrictive growth management policy and zoning referenda become in the suburbs, and how fast Cleveland empties of its population and institutions that are beyond their useable life may be beyond our influence. I think a lot of redevelopment will have to be done so that we can acquire abandoned land, clear the title, package it, and merchandise it in a market that is looking for ripe land. Now if the Cleveland area as a whole continues to decline, it is hard to imagine where the growth will come from to rebuild. It is not hard to imagine, however, other uses for the land that are less urbanized in nature. It is very, very difficult to talk about this sort of thing in cities like Cleveland where many politicians cannot even bring themselves to say that the population is really declining. It is a serious problem for a mayor, for example, to say that the population in Cleveland has gone from 914,000 in 1950 to about 690,000 now. Yet there are politicians here who cannot say we are losing population even after seeing the numbers. They cannot say we are losing population because their opponent in the next election will say, "It's your fault!" So we go about making believe that we have no problem. Similarly it is very difficult to talk about cities like Cleveland—which are major urbanized, heavy production, durable goods-producing cities—becoming semi-rural, and yet in neighborhoods like West Central and Hough you can look through the blocks. We have densities in those neighborhoods, formerly teeming slum neighborhoods, that are as low as Pepper Pike, which is very fashionable suburban area. Nobody is there any more. Whether somebody will consume that for urbanized uses or not, nobody can tell. But surely that land can be used for something more amenable, maybe a regional park, farmland, or who knows.

Q: I would like to get a little deeper into this Muni-light

question. Why did you want to condemn the private company? Because it was ripping off the consumer or because the municipal publicly owned facility was struggling and they needed the extra customers to bring it up? If so, why do you think that the public sector could do a better job than the private sector?

R: That is a long question, but it is a good question. Basically, in Tom Johnson's Administration (which was around the turn of the century), the reformist administration set up Muni-light over the objections of Cleveland Electric Illuminating. Cleveland Electric Illuminating has tried since 1908 to put Muni-light out of business. It has a very well-documented history of doing everything it possibly could do to drive the public power competitor into the wall. I think that is a reasonable thing for a business to do, so I am not surprised at that. At the same time, Ohio law gives us the power, public power, to condemn, in the public interest, any investor-owned utility in the state. Because Cleveland Electric Illuminating was engaging in restrictive practices, Muni-light could not get a connect or an interconnect with any of the regional power grids (like New York or the Tennessee Valley Authority), and so when our equipment was down, we had to generate beyond that equipment and suffered outages and that kind of thing. A normal power company simply throws a switch and connects to New York or TVA. Cleveland Electric Illuminating refused to let us use their power for emergency purposes. Muni was in bad shape; it had a series of outages and was teetering on the edge of extinction for many years. Frankly, I was outraged with the tactics Cleveland Electric Illuminating was using very successfully and was confident that Muni could be saved. Certainly with Cleveland Electric Illuminating production facilities, the city could produce electric power at a much lower cost to the people of Cleveland. Now that only makes sense. You know there is more than a personal vendetta involved. It is more than a personal sense of outrage. If cities like Cleveland are to be in an attractive location for residential and industrial development, then they have to have something to offer, something competitive to offer. It seems to me that if the City of Cleveland could offer low-priced electric power to a footloose industry, maybe we would have the means of attracting some other industry. It would bring strength to our competitive position. We must improve our competitive position.

QUESTIONER'S RESPONSE: Lord knows you are trying!

REFERENCES

Jerome Kaufman, "Contemporary Planning Practice: State of the Art," in David R. Godschalk (ed.), *Planning in America: Learning from Turbulence* (Washington, D.C.: American Institute of Planners, 1974).

Paul Davidoff, "Advocacy and Pluralism in Planning," *Journal of the American Institute of Planners* XXXI (November 1965).

Chapter 4

INDIANAPOLIS: FRAGMENTATION AND CONSOLIDATION

MICHAEL A. CARROLL, AIP
Deputy Mayor
Indianapolis, Indiana

I would like to give all of you the opportunity to understand the political, social, and economic aspects of Indianapolis and Marion County, in central Indiana, so that you will have a feel for the nature of planning and the public decision-making process. The Central Indiana Metropolitan Statistical Area is approximately 1.2 million in population. There are seven counties, including Marion County, in this area. The development pattern is a result of the urbanization of Indianapolis and Marion County, urbanization that has extended beyond Marion County into both the northern and southern portions of the Central Indiana area. The population of Marion County is about 800,000. The racial distribution is approximately 81% white and 19% black in the entire county. Within the old corporate limits of the City of Indianapolis, which at one time made up about one-third of the entire county, the racial distribution is approximately 30% black and 70% white. The population of the old corporate limits of Indianapolis amounted to close to 500,000 before city/county consolidation of Indianapolis and Marion County.

In 1969-1970, the Indiana General Assembly passed the

Consolidated Cities Act of 1969. *That particular act consolidated both the city and the county government in terms of political jurisdiction of Indianapolis and Marion County.* It did not automatically change all of the taxing units that existed within the county. We still have about 74 separate taxing districts for various functions. What it did do in the major sense of the word was consolidate the legislative body. There used to be a City Council for the City of Indianapolis (approximately nine members elected at large) and a County Council for Marion County (five members elected at large). This body was consolidated into one countywide City/County Council of fourteen members, and over a period of two years it made the transition from all at-large elected representatives sitting as a Transition Council to a 29-member Council: 25 elected from districts of about 31,000 population each and four at large.

This major consolidation of the legislative process in the county was complemented by a major consolidation of the executive or the administrative branch of city/county government. The Mayor, formerly the Mayor of the City of Indianapolis, became the Mayor of Marion County or, in effect, the County Chief Executive (with the exception of four smaller communities amounting to approximately 45,000 population within the county). The Mayor is elected by all of the citizens from within Marion County and is also responsible for the appointment of departmental directors to six of the major departments of this Unified Government or "UniGov" as it is sometimes called. The Department of Metropolitan Development, of which I formerly was the director, has the responsibility of comprehensive planning and zoning, in addition to public housing, urban renewal, buildings, code enforcement, and preparing our community development applications. Five other operational departments are Administration, Parks and Recreation, Public Safety, Transportation, and Public Works. Those departments are representative of what, prior to consolidation, were 58 separate units of city and county government. Now consolidated into six major departments, with all directors appointed by the Mayor, it is a strong-mayor form of government.

There are two Deputy Mayors in this particular form of government. I am basically responsible for the elements of metropolitan development, downtown revitalization, capital improvements programming, and policy coordination of all federal programs from the level of the chief executive. The other Deputy Mayor is basically responsible for liaison between the Mayor's Office and the City/County Council; for interaction with the Indiana General Assembly, in terms of our legislative programs; and also for political liaison. This is a general description of the nature of our local governmental structure from the standpoints of the legislative body and the chief executive.

We had a planning process in Indianapolis, prior to Marion County and Indianapolis consolidation, that was initiated in 1955. In 1955 Indianapolis and Marion County joined in the Indiana General Assembly to send through a piece of legislation that created one countywide planning department with responsibility for capital improvement programming, zoning, master planning, subdivision regulations, and basic controls over the implementation aspects of planning on a countywide basis. So as far back as 1955 in our community we had countywide comprehensive master planning. After consolidation, that planning department became the Division of Planning and Zoning within the new Department of Metropolitan Development. Planning moved from a staff function to a line function within an operational department of local government. From 1955 through the period of consolidation, the nature of planning in Indianapolis and Marion County changed substantially. The planning done in Marion County from 1955 to approximately the mid-1960s was basically physical in nature.

As the new planning department began its work, most of the major emphasis was on countywide land-use planning and attempting to pull together the various facility components of the comprehensive plan in terms of utilities, services, schools, parks, and so on. In addition to that, moving into the mid-1960s we began to get into areas of economic planning. To some extent, we dealt with social planning as it related to neighborhood revitalization and as it tied into some very serious problems in the inner city, problems resulting from major relocation occurring in the inner loop of our downtown area. This began to change the nature of the planning process in Indianapolis and moved it more and more in the direction of responding to the difficulties of neighborhood revitalization; relocating of the people involved in major governmental action projects; and developing a planning strategy for achieving closer involvement with the decision-making process.

Our unified planning work program is fairly comprehensive. After we consolidated government in 1969 and 1970, we moved into what we called locally an Integrated Grant Administration Program for planning. It began to tie together all of the federally funded planning programs that we had some access to and led to involvement with one grant application, one form of administration, and one form of management control in terms of our relationships with the functional departments in local government. We are one of four major communities throughout the United States that initiated the Integrated Grant Administration Program which found its way into federal legislation in 1974 through the Joint Funding Simplification Act. The benefits of this approach to planning are the integration of all aspects of comprehensive planning into one work program; the tying in of various funding sources to that same work program; and one form of managerial control in a centralized planning department within a major arm of government.

We moved from that particular point in 1969-1970 to a phase of social planning that was an outgrowth of our involvement in the Model Cities Program. The City of Indianapolis Model Cities Program began in 1969-1970, in terms of its planning year, and has recently completed its fifth year of implementation. It has now been folded into the Community Development Block Grant Program. We began to recognize in the early 1970s that we had no major social components in our comprehensive planning process at the citywide level, and the Model Cities Program required such a component for dealing with the problems of neighborhoods: crime and public safety, health services, day care, unemployment, and welfare difficulties. We formerly had not addressed these kinds of neighborhood problems through planning. That began to change in our metropolitan planning program; we began to do some citywide or countywide planning in the area of social program activities.

We began to expand our approach to comprehensive planning, and we moved through the mid-1970s into a period that most cities have faced for the last four to five years—a series of economic and fiscal difficulties. The change in the nature of our planning program has reflected our concern with these difficulties. We have been engaged for the last three years in a series of work programs and planning studies that tie into the questions of bonded indebtedness in our community, the capital improvement programming process, and the city budgeting process. In effect we moved from a primarily physical planning emphasis in the late 1950s and early 1960s to a neighborhood and social focus in the mid-1960s and early 1970s, and, in the mid 1970s, *we are concentrating substantially not only on the integration of these parts of the planning program on a citywide basis, but also on an attempt to make planning more relevant to the decisions made in the budgeting process.* This is currently the major focus of our planning program in Indianapolis and Marion County.

We still do physical planning and have a countywide comprehensive land-use plan that was adopted in 1969 and that has subsequently been revised and modified by subareas throughout the entire county. We have done physical planning by attempting to relate it to major private investment decisions in Indianapolis and Marion County. We moved from a countywide level of planning to a series of major planning subareas that make a major difference in the quality of life in Indianapolis. A major product of this effort has been our Central Business District or Regional Center Plan for Indianapolis. Our major focus on downtown is representative of a concern that we have had for a long time. The current plan represents the sixth revision of the downtown plan within the last fifteen years. It gives an idea of how rapidly both economic and institutional conditions change in one particular area of the community. It requires constant monitoring and updating in terms of short-range changes in the planning philosophy.

The largest private enterprise downtown renewal project that we have engaged in is called Merchants Plaza. It has a 520-room Hyatt Regency Hotel, which is directly adjacent to the Indiana Convention Exposition Center built in 1968 to attract major conventions to Indianapolis and to the State of Indiana. The City of Indianapolis was involved directly in that particular project through planning and urban renewal by actually selling a redevelopment bond issue to acquire the ground and to remove the dilapidated structures. This paved the way for a bid that ended up in a lease arrangement, between the city and the developer, to build the Merchants Plaza. In addition to the Hyatt Regency Hotel, it includes several major downtown office buildings. This is an example of the implementation of planning that I think we can point to in Indianapolis with some pride. With the use of not only public financing, but also public involvement in downtown planning, the result was a major private investment to the benefit of the total community. Our focus on downtown is clearly seen in such plan implementation.

Another example of plan implementation downtown is Market Square Arena. While the domed exposition center was financed by public funds, the parking garages north and south of the structure were financed privately. The project includes an important restoration of the City Market directly adjacent to the City/County Building. It was related to other private investments north of the City Market that includes an office building and parking garage. The entire project is financed at the level of $43 million, of which approximately $22 million is from public funds. The restoration of our City Market is an interesting project. It was originally created by the City of Indianapolis in 1850. It burned down in 1888 and was rebuilt on the same site. It became a combination of what is called Tomlinson Hall, the old city hall, and an open-air market. We initiated the project with a $4.7 million grant from the Lilly Endowment, a major private foundation in Indianapolis. The endowment made this project their first priority in 1975-1976. The central structure of the City Market is on the National Historic Register, and we are in the process of restoring it first. We have planned an adaptive restoration in order to retain many stand merchants and market people who were thinking of moving before. On the east and west sides of the City Market we will have an urban park. Next to the City Market is a new office building and parking garage which was recently constructed and is about 80% occupied. This is the kind of integrated development that is possible with public and private joint ventures.

Still another example of our downtown focus is our plan for the Lockerbie Square Historic Preservation Project in the northeast downtown area. This plan was prepared over a number of years to complement a Preservation Zoning Ordinance. The ordinance and plan are attempts to preserve the physical character of the neighborhood that was the subject of James Whitcomb Riley's early work in Indianapolis. His home has been preserved, as have a number of the residential structures in that neighborhood. The city has participated by assisting with a cobblestone street restoration and partial restoration of historic structures. There has been over $1.5 million worth of new investment here; much of it comes from suburbanites moving back to the Lockerbie Square Area. They buy usable structures and restore them in a manner consistent with the theme of the project.

Directly north and east of the downtown area is the Model Cities Area, for which we have completed a land-use plan. We took a portion of this area that was severely deteriorated and dilapidated and acquired (through the Open Space Land Grant Program) twelve acres for Martin Luther King Park, the first manifestation of the plan. We then designed a sixteen-block plan called Park II to protect the park as well as to begin local urban renewal activity and encourage major public investments in this worst part of our deteriorating inner city. At the southern portion of this area we identified locations for a new health center, a school, some new private investment in terms of inner-city housing that we built there, and a $1.3 million Ambulatory Health Care Center adjacent to the Martin Luther King Park. With the demise of the Federal Model Cities Program, we now use our Community Development Block Grant to finance the basic operation of that health center at the level of $500,000 per year. There is also some new private investment of approximately $1.5 million in the Caravelle Commons, which is a Section 236 housing project adjacent to the health center.

Our basic approach for the last six to eight years has been (1) to tie planning specifically and directly into the process of encouraging private investment with public money, and (2) to tie planning into the decision-making and implementation process in local government. I think we have been successful in Indianapolis in attempting to initiate this kind of process. It has not been without pain, because our planners are required to deal on a day-to-day basis with the entire bureaucracy, which remains fragmented even with consolidated government. This is one of the major problems that has to be resolved in any metropolitan area. At the least we need, to some extent, a reduction of the tension between those who operate in the field of planning and those who operate in the day-to-day decision-making process. We have been fortunate in both former Mayor Lugar and current Mayor Hudnut's selections of professional department heads. They have chosen people who have professional backgrounds and experience, not only in local government but also in the private sector.

We still have 29 City/County Council members who must be involved in the planning process and the subsequent appropriation of public funds to implement programs. This is where planners who aspire to implement planning must really work on a day-to-day basis. It is only through the political process of developing trade-offs and compromises that we can affect the decision-making process (and attempt to translate that into fifteen votes on the City/County Council). This is a very important process that some planners try to ignore. It speaks not only to the technical and methodological capabilities of those who do planning, but also to the importance of communication among those people who have political clout—in the sense that fifteen votes are necessary to get something through the City/County Council. There are people who surround the chief executive who can provide a communication link between planners and there are people who can tell you what is necessary to get a decision made.

As one moves through the administrative and managerial structure of either local government or state government, one who comes from a planning background, such as myself, finds how very important it is to begin to develop a personal credibility in the decision-making process. As an individual, you have to be trusted by the people you deal with. The recommendations you bring to them will then have some credibility because they have been tested in the community. They must be tested by virtue of your exposure to citizen groups and the City/County Council, by testifying before the State Legislature, and by your involvement with members of the private sector. If you do not retain the confidence and trust of those people, nothing you do, no matter how technically superior, will ever become reality. It is very important not only to maintain that trust, but also to understand and work within the political process. You should not be afraid of the political process.

Politicians must have confidence both that you will be accountable for proposed expenditures of public funds and that you will be around long enough to see that such plans will be implemented. You subsequently will be called on the table (three or four or five years from now) to answer why something went wrong, or why what you said five years ago is no longer relevant, or why the plan was not implemented as you had proposed it to the City/County Council. Those are the critical questions that you have to answer as a professional planner.

In summary, it is very important for an individual with technical and professional capabilities in planning to begin to branch out and understand—as you are exposed to the operating realities of day-to-day government—what the decision-making process is and what the parameters and constraints of that process might be as it relates to operating department heads. Then try to reach an accommodation between the question of long-range planning, which may not have any relevance to day-to-day decision-making, and short-range functional planning, which does have relevance to day-to-day decision-making. You must also determine the relevance of allocation of public funds, preparation of the annual budget, and capital improvement programs for long- and short-range planning. As planners in local government, we have many tools to work with in terms of implementing planning and making it effective. It is important to develop a method, style, and philosophy for using those tools and generating the confidence of those that you deal with in terms of educating them about the use of these tools. Why capital improvement programming is important should not be stated from a planning standpoint but rather from the standpoint of better allocating local resources in the time of fiscal scarcity. Deal with the decision-making process on its own ground. Do not attempt to argue to implement a decision on the basis of the planning rationale behind the decision. Argue on the basis of the economics of that decision. Argue on the basis of the political sensibility of that decision. Argue on the basis of the fact that it is going to be good for the community. Argue that you have attempted to define the public interests in such a way that you touched base with all of the groups that are likely to be affected by the decision, and argue that this is the best compromise you can arrive at with the least negative consequences. Those are all very important parameters for the decision-making process. That is how I see a professional planner in local government with the opportunity to make a real difference in terms of the decision-making process.

That brings us inevitably to the applicability of planning theory. Well, the only theory I know of that we operate with in the City of Indianapolis is that we think planning is comprehensive—which includes social, physical, and economic components. We think planning should be long-range, but it is not. It is short-range. We think planning should relate to the day-to-day operating requirements of the departments that we deal with. We think planning should tie into the budgetary process. We think planning should be sensitive to political and community values that relate to the area within which we operate. That is the only applicable theory I know. All else is academic.

QUESTION: If you were to summarize an overriding goal or objective for the planning and management processes in Indianapolis, what would it be?

RESPONSE: I would say, as a planner and public administrator, that the most important goal for the decision-making processes in local government is a combination of two things. First would be *input into the goals-setting process of the political administration within which one is operating.* That means assisting the Mayor in preparing the state-of-the-city address; assisting the Mayor in preparing the budget message; attempting to develop consensus and then translate that consensus into decision-making; and participating in the goal and objective-setting process in the community in the way that those decisions are made. That is how to meet such an overriding objective.

There are goals studies that many communities have done to try to involve lots of people in the community to generate a consensus. They ask where that community is going, and what its basic objectives might be, and somehow translate the responses into political decisions. In other communities this kind of process does not take place. In Chicago, Indianapolis, and other cities, decisions that are made relative to the city administration are made by the Mayor. Those decisions are reflected in public statements of policy that the Mayor makes at the beginning of each year, setting out his major objectives for that year. It may be economic development; it may be economic base revitalization of the downtown; it may be a series of these things; but input into that political process is very important to what comes out.

Second, I think *the budget-making process is the most important public decision that is made at the local level at any time during the year.* If in fact anything is going to happen, in terms of local government expenditure and allocation of local resources, it happens through the budget-making process. If a planner has input into the process of preparing a capital improvement program that is adopted by the Council—which is the basis on which the Council then allows bonds to be sold or tax revenues to be generated for specific projects—then the planner has made a difference in that community. One makes a difference in the operating budget by making sure that there is a relationship between capital expenditures and operating expenditures and that these tie back into the Mayor's goals established at the beginning of the year. In that way the planning process supports the nature of decision-making on an annual basis as well as on a short-range basis over a three- or a four-year period.

Q: Most of the projects you discussed were funded under the old categorical programs. What kinds of changes have you seen in your own local planning operations with the advent of revenue-sharing, and what kinds of changes would you like to see in the revenue-sharing legislation?

R: I think a major change we have seen in the City of

Indianapolis is the movement in 1972-1973 to a form of generalized funding from the federal government. General Revenue-Sharing, which has very few strings attached, has eight priority areas for expenditures for local units of government. Local councils and mayors can use that particular source of revenue to fund one-time capital projects or to fund deficits in operating departments. It has become the glue that holds the local budget together. At least that was and still is the case in our community.

The movement from categorical grant programs has been in the form of what we would call Special Revenue-Sharing or Block Grants, such as Community Development, Manpower and Job Training, Criminal Justice, and Health Service Related Programs. The implications that these kinds of changes have had are very serious ones in terms of how decisions are made in local government. In many cases they have caused a restructuring of local government because of consolidation into one function of what were formerly separate functions. In order to respond better to the decision-making process at the local level, some governments have reorganized themselves so that one department, for example, would be responsible for the planning, development, and expenditure of the Community Development Grant. The same would apply for the other block grant areas. These changes have caused questions to be raised at the level of the city council about the budgeting process and allocation of resources. Where resource allocation used to be hidden or fragmented among a variety of different programs, in many cases with overlapping responsibilities, now the total expenditure for a function is exposed in the political process. One can really see the true percentage of funding spent on a particular component of local government vis-à-vis other priority areas of expenditure. I think it has been easier to administer, manage, and make decisions about the funds that go into block grant programs than it ever was before because of the reduction of fragmentation and better consolidation of effort.

I think in addition that it has made urban management easier from the standpoint of the executive decision-making process, management, and implementation of the services that are delivered with those funds than ever existed before. I would say that those are probably the major changes that have occurred, at least in most local governments that I am aware of, as a result of the movement from categorical grant programs to general-revenue sharing and block grant programs.

The new General Revenue-Sharing legislation has two major changes in it that are very significant to local government. One is that the General Revenue-Sharing money can be used to match other federal grant programs, which was not the case before, so the leverage we now have with General Revenue-Sharing money, even though the dollars remain the same, is multiplied substantially. As a result of criticisms in the first five years of General Revenue-Sharing about the lack of public input, there have been specific requirements incorporated in the law for citizen participation and public hearings to take place that tie into the budgetary process. Again I think this is an improvement in the nature of public decision-making. We have seen as a result of the movement to block grants (and we moved in this direction in Indianapolis in 1971-1972) an incorporation and an appropriation of all of the funds that local government deals with on an annual basis. I think it has tended to solidify and make more comprehensive the budgetary process because we appropriate all of our taxes and revenues from local, state, and federal sources at one time, that is, during the city budget process. We used to only appropriate local money and not have to appropriate the federal money, but now we have a gross budgeting procedure that allows us to relate the federal money to the state money to the local money and tie that in to local priorities.

Q: What were the effective political ingredients that made possible, encouraged, and promoted the consolidation of government in Indianapolis?

R: It would be unfair to say that there was any one particular reason that stands out as we began to move in Indianapolis and Marion County toward countywide functions in the early 1950s. When people perceived in an incremental fashion that health services could no longer be financed on the tax base that existed in the City of Indianapolis, and that health problems existed in the county, we had to go to the General Assembly over a period of twenty years to expand functions from the City of Indianapolis to Marion County. We expanded the health function in 1953, the planning function in 1955, the parks and recreation function in 1961-1962, and the transportation function in 1964. A Metropolitan Government Bill was initiated by the predominantly democratic city administration in 1965 in the Indiana General Assembly—it did not pass. In 1967 there was a Metropolitan Police Bill prepared that would have consolidated the Police Department and Sheriff Department on a countywide basis. That did not pass.

Mayor Lugar ran on a platform in 1967 that had governmental reorganization as one of its major priorities. He recognized very early that being Mayor of the old City of Indianapolis was being Mayor of little or nothing because the Mayor had little or no control over the functions and services. He put together a task force that initiated the draft legislation. He campaigned throughout the state for people who were running for office in the Indiana General Assembly that year. That year happened to be a Republican year in Indianapolis, Marion County, and Indiana; Mayor Lugar was elected, a Republican governor was elected, all of our delegation for Marion County was Republican, and all of these people had agreed to the platform that was prepared for governmental reorganization. It passed in the Indiana General Assembly in 1969, and it passed with substantial community support. The Mayor ran again for office in 1971, and the biggest issue in the election was the "worthwhileness" of UniGov. This was really a referendum on governmental consolidation as part of the mayoral race. Lugar won by 51,000 votes, primarily because the promises that were made for UniGov have been carried out. We have not raised the city tax rate. From 1969 through 1976, the tax rate was the same for the same functions that were performed in 1969, but more services were delivered. They were delivered with more efficiency and rapidity than they were before. I think the economics of the measure proved to be successful. Also the practical aspects of implementing the governmental restructuring did not prove to be onerous in terms of traditional tax requirements on the citizens of the community.

Q: Have there been any problems with the unification?

R: I would say that the problems that have occurred were the normal ones that would occur in any transition from a fragmented governmental structure to a streamlined management operation. We had a lot of people who probably were incompetent, and many were performing functions that others were performing in the city. There was a large amount of struggling for power in the first two years of the unified governmental structure, and I think this probably would have occurred in a governmental reorganization at any level. There were problems of loopholes that were left in the original legislation that we had to go back to the Indiana General Assembly to modify. Each year we go to them with a UniGov Omnibus Amendments Bill to tighten up some loopholes, to continue to reduce duplications in various portions of the structure, and to centralize other kinds of functions.

Unfortunately, we are in a position where, if we want to create a new Department of Local Government, we have to go to the General Assembly of the State of Indiana. We are not a home rule state, and we do not have charter prerogatives. In the State of Indiana, any modification in the structure of local government requires approval by the Indiana General Assembly. This has created substantial problems for us in terms of the flexibility we would like to have.

Q: What then were the major problems of the consolidation?

R: The major problems I would say were (1) replacing people who were no longer able to accept substantial performance responsibilities under the reorganized government; (2) obtaining the appropriate understanding on the part of the citizens in the community and in the private sector as to how the new government would operate; and (3) having the departments become familiar with each other in terms of functions, responsibilities, and working relationships.

Q: In your discussion of functions unified through consolidation, you have failed to discuss education. How has it affected education, usually one of the more costly public services provided by local government?

R: First, under the state constitution it was impossible even to consider consolidating schools. Second, it was politically inappropriate because the bill would have never passed if school consolidation had been part of that particular piece of legislation. We still have eleven separate school corporations within the boundaries of Marion County, the largest one being the Indianapolis public school district, which is basically the old corporate limits of Indianapolis before consolidation. There are eight township school corporations within Marion County but outside the Indianapolis public school district and two school corporations in the excluded cities. So the decision-making process is separate, with each school corporation having an elected school board that sets its own tax rate and basically does not participate in financial planning or programming with the general purpose government.

It is correct that, if we look at the total tax bill, schools take anywhere between 45% and 55% of the total tax bill. We are engaged in extensive litigation in terms of school desegregation, with proposed remedies that have ranged all the way from consolidation of all of these eleven separate school corporations to including even townships outside of the county. Now we are down to a form of intradistrict busing, with the case currently before the United States Supreme Court. But basically our school districts remain separate governments, separate taxing units, and do not participate in the general decision-making process of local government.

Q: It would be fair then to say that consolidation really has not addressed the problem of property rich versus property poor school districts or questions of black-white ratios in various schools districts.

R: No, and it did not purport to do so. It would be unfair to say that it could be measured on the basis of that criterion.

Q: Would you tell us a little bit about the so-called MiniGov proposals and what the status of those proposals is now?

R: MiniGov was initiated in 1969 when the UniGov legislation was going through the General Assembly. There were opponents to the City/County Consolidation in the Indiana General Assembly who felt that a centralization of power in the hands of an elected County Chief Executive would be too strong a form of government. They feared it would not have the checks and balances that they felt to be appropriate. The argument began with, "Should we have eleven councilmen all elected at large to counterbalance the power of the executive branch of government, or should we have 50 councilmen all from districts to put a decentralized focus on the decision-making process?" What developed out of that was a compromise between eleven and fifty that ended up being 29 councilmen: 25 elected from districts, four elected at large. The nature of the decision-making process at the legislative level really changed in terms of perspective, but part of that companion bill was that there was some trade-off in the Legislature. One of the Marion County representatives had proposed MiniGov legislation that would have created what we would call Community Councils throughout Marion

County. The bill did finally pass in 1972, and there was an attempt to have it repealed in 1973 by supporters of the City/County Council recently elected to districts. In 1974 the Council chose not to approve the plan that they were legislatively required to approve by MiniGov. That effectively killed implementation of the bill.

The basic concept was to set up a system of neighborhood governments. The Metropolitan Development Department was required to propose districts for the entire county on the basis of a series of typical neighborhood delineation criteria. I was Director of the Department of Metropolitan Development at that time. The year after the bill passed, I was responsible for presenting the Council a district plan for the entire county. The Council never could reach an agreement on districts. They considered themselves as already being representatives of such neighborhoods. So I would like to look at MiniGov in this vein; it was probably an idea whose time had not yet come—at least for our political situation.

Q: You mentioned several times the importance of the link between the planning process and budgeting. I think it would be interesting to hear some more detail about how that link is institutionalized in Indianapolis.

R: I think that link was institutionalized by virtue of some efforts that initiated out of our Model Cities Planned Variations Program and our 701 Planning Program. We dealt with the questions of building a management capacity in local government. Even though we had unified government in 1970-1971, we did not have a unified budgeting process. We were still appropriating only local money. We were not doing any form of program budgeting. We were not doing any capital budgeting. There was very little rationality in the process by which budgets were made at both the executive and legislative levels. We began pushing for the institution of capital improvement programming and budgeting because that was something that could be seen, felt, and touched. It was a physical representation of facilities and networks of systems in the community that made a difference in terms of growth and development.

I think from an educational standpoint, we initiated that process in 1972. In the first year it was only the collection of an inventory of anticipated capital expenditures by the various departments pulled together for the first time. Everyone could see what everyone else was doing, with no policy input or attempt to evaluate or make any policy judgments about where growth ought to be or where it should not be. We attempted to visit with the Council to explain this process to them. In the second year we got the Council to pass an ordinance requiring capital improvement programming and establishing a time framework that would tie the preparation of the capital budget into the operating budget cycle and process.

We spent a substantial amount of our Model Cities money, which came out of a planning program initiated in the Planning Department, to institute program budgeting in local government. It took a period of three years to institutionalize the system. It was a computerized software package—from Peat, Marwick, and Mitchell—called the Financial Accounting Management Information System. It was a computerized system of classifying budgetary expenditures by program, sub-item, and activity. Then it ties that all together in terms of a summary budget by playing out all of the state, local, and federal revenues, allowing one to see where the money is going. Once the system was in place, with the computer up and running and the program debugged, the Council understood what was going on. It really was not until the 1975 budgeting process that the Council members felt familiar with that system for making decisions about the budget. Even then the Council was concerned that things were being hidden by the program budgeting process and aggregation of expenditures. "You were hiding really how much that guy was making in the planning department," and "you were hiding a number of other expenditures that councilmen are traditionally concerned about." Therefore the year before last year we had a dual budgeting system. The Council members felt comfortable, and they could relate to the old system of decision-making. After that year of preparing a dual budget, we went with the program budget. The computerized form of budgeting became familiar to the Council.

It makes good sense to me to institutionalize and link major expenditures in local government with planning, but it cannot be done effectively unless one is willing to commit four to five years. The institutionalizing process requires educating the legislators, department heads, and clerks, the people who do the accounting, bookkeeping, and classifying of expenditures by sub-activity, program, and everything else. It takes a long time to get that decision-making process included in the daily thinking of people who do budgeting.

Q: How does the City of Indianapolis view the roles of the automobile and mass transit in its future development?

R: We have done the traditional long-range transportation planning and updating of our forecasts in terms of modal splits between mass transit anticipated trips and vehicular trips, and we have investigated both how those would be distributed over our interstate network and our thoroughfare network and how that would be related to traffic generation points within the metropolitan area. Basically we have seen Indianapolis as a very low-density community. Growth is not what you would call "sprawl," but it is moving out in suburban direction because land is an available commodity. In addition to that there is still a large amount of agricultural land within Marion County, and, even though the price of that land is going up, high-density development is not the order of the day. In effect, the ability to channel development into corridors of high density to make mass transit economical is minimal.

At this point what we have in our transportation plan are park-and-ride facilities, using an express bus lane on the interstate. We have a transit mall that we are attempting to implement downtown, but there are some other things that have to happen first. We have applied for a downtown people mover project to demonstrate pedestrian circulation at the elevated second level. We are trying to deal with the implementation of a second level pedestrian network; this began with our Market Square Project and tied into the City Market.

Right now people are still tuned in to the automobile. We do not anticipate that development patterns in Indianapolis will generate, in any short-range period, the economics necessary to make mass rapid transit feasible. What we are trying to do is make our bus system more flexible, rescheduling and rerouting it to market demands in such a way that we can capture a latent demand. I do not see any long-term change in the access of vehicular traffic, particularly in the downtown area. There is a great deal of resistance from the business community; these people feel very strongly that without access by automobile there would not be people coming downtown. Our inner loop highway provides the accessibility that downtown has not had for a long time, and that accessibility is precipitating an increase in land values. It has generated private investment for distribution warehousing in the downtown area, and has brought people downtown to those major public generators that we have invested in—the arena, convention center, and Merchants Plaza. Those people come downtown by automobile, so as long as that kind of habit and the economics of the system continue to work, we are going to accommodate that kind of traffic to the best of our ability. Both the public and private sectors have a vested interest in automobile accessibility to the downtown.

Q: Of the people that work downtown, what percentage ride the buses?

R: I would say no more than 5% to 10%, at the most. The overall ridership in the metropolitan area is less than 3% in terms of all the trips taken. Indianapolis is an automobile town.

QUESTIONER'S RESPONSE: We know!

Chapter 5

KANSAS CITY: PROBLEMS AND SUCCESSES OF DOWNTOWN DEVELOPMENT

JOSEPH E. VITT, JR., AIP
*Director, City Development Department,
Kansas City, Missouri*

I would like to summarize what I intend to do. First, I will describe the background of planning in Kansas City. Second, I will describe my approach to managing the department—the theory if you will. Third, I will briefly describe the management plan of the department—i.e., the basic operating procedure for the department on a day-to-day basis. Fourth, I will describe briefly some of the department's programs that might be of some interest.

The planning program in Kansas City began much later than in many other middle-sized central cities. Perry Cookingham, who is often referred to as the dean of city management in the United States, issued a report on planning for Kansas City (which is also an American Society of Planning Officials document). The planners felt that it was such a good basis and rationale for beginning planning in the city that they issued it in the late 1940s as a publication for planners and planning agencies. Cookingham started the program in about 1943; that was pretty late for comprehensive planning to begin. He hired a few staff members because the Council approved that plan. That led to the development of a master plan for the City of Kansas City in 1947. That plan was for a much smaller area than Kansas City covers today. It was probably for an area of 70 or 80 square miles, while the City of Kansas City today is 320 square miles, the result of a number of annexations during the 1950s and 1960s.

The city planning program was undistinguished through the 1950s and 1960s—nothing very much happened. I think a search through the literature probably would not find anything of much importance on planning in Kansas City. It was not helped by a change in administration during the period 1959 to 1963 with Democratic factions as a holdover from the old Prendergast Era. The Prendergast Machine in Kansas City, which some of you may have heard of, took office for that four-year period, and it was known as "the reign of terror." City managers came and went at about the rate of one every six months; most of the professional staff left for other jobs; and it had a devastating effect on what there was of the planning program. The city never really recovered during the rest of the 1960s.

In 1968, the City Planning Department was renamed the City Development Department. The main change was to add some social and economic dimensions to the department's physical orientation. It had basically been producing physical planning, land use, zoning, subdivision regulations, and not much else. So they thought that by changing the name of the department and by adding a new function, they might be able to change a little bit the orientation and, perhaps, improve the quality of the products coming out of it.

Until 1972 the department was plagued by a number of factors that limited its effectiveness. There were many changes in directors; the longest any director ever stayed was about a year and a half, and most of them were not people with planning degrees; rather, they were public administration people. They were actually moving on to other jobs, but they happened to be heading this department while they were in a holding pattern. At least, it seemed like that to the staff. Second, there were tremendous pressures for redevelopment reviews because there was a great deal of land added through annexation. This was a period in the United States when a lot of development activity was taking place. In Kansas City many developers were applying for rezonings. The small staff the city had was working on rezoning cases, and that left little time for addressing the much larger development issues that the city faced as the result of both annexation and the aging of the central city. Third, there was an inability to attract the quality staff needed to deal with these problems of rezoning and growth pressures, and, at the same time, to try to put together a comprehensive plan. There was no change in the comprehensive plan—no effort had been made to update it since 1947—so when I came in at the end of 1972, we had a staff that had not really done very much with planning and certainly had not participated in any updating of the comprehensive plan. There were some changes taking place with respect to social and economic planning, and there was some interest in housing conservation, but not much more than that.

My approach to directing the department is the result first of all of my education and experience, particularly my experience in the Mayor's Office in Detroit, where I spent about three years as an Executive Assistant to the Mayor for Planning and Development (basically the Development Coordinator for the city). This is where I was able to help apply management principles and concepts to the management of a large city. The Mayor of Detroit, Mayor Gribbs at that time, operated the Mayor's Office as a management organization. He felt his job was to manage the organization and not to be very political about it, to the chagrin of his political advisors who felt that he was not doing the right thing by trying to manage resources rather than looking out for political opportunities. So the staff was charged with managing a 23,000-member, $7 million per year organization, which is the City Government of Detroit.

I learned a lot about planning and management, what planners do, and how they are viewed. I had to manage seven departments, including, the City Planning Department, Community Renewal Program, Industrial Development Commission, and a number of other physical-development-related organizations. This is where I became convinced that planning agencies and planners could become more effective if the best concepts of the management profession—that is, things that industrial managers and business managers were telling us and were using in their own professions—and the basic concepts of the planning profession could be combined, integrated within an organization, and then given some time to develop so that we could see the result of that management organization and that management plan.

The Department of City Development in Kansas City has about 130 people. We are responsible for planning and programming of physical facilities, housing and economic development, the capital budget program, the major street plan, and the parking and boulevard plan. We are also the zoning and subdivision review agency. We are the review agency for the 353 Program, which is the Missouri Redevelopment Law that allows private developers to use the city's powers of eminent domain and tax abatement to redevelop blighted areas. We are also the agency that reviews the planned industrial expansion projects that provide financing—revenue bond financing—for industrial development projects. So we are the agency that is most involved with planning and programming for physical development within city government on a day-to-day basis. We have a budget of about $1 million for the staff, and we directly administer about $6 to $7 million in additional funds under the Community Development Program and other programs. We of course program for projects like the Community Development Program, which in Kansas City, on a yearly basis, is about $15 to $17 million dollars a year. So we are heavily involved in planning and programming in all of those areas.

The basic approach that I have taken to the development of the department is a management-oriented approach. It is distinguished by a number of characteristics. First of all, it is an integration of the planning function and action into a single, continuous, management process. This management process includes five closely related steps: (1) deciding what to do; (2) committing ourselves to do it; (3) controlling the doing of it; (4) evaluating what we did; and (5) feeding the evaluation back in to decide what we want to do the next time. In this model, which we try to follow in the department, planning is not separated from action; planning is an integral part of the management process. We provide management services to a whole range of clients who do not provide planning services. We do a lot of planning, but we are basically helping groups of various kinds to manage their resources. We provide planning as part of that, but we also provide implementation. Working with the Council, for example, we get plans approved and find resources. So we are involved in a management process where planning and action are tied together.

Another characteristic of our approach is the focus on helping make decisions today that create a future tomorrow. We are not really interested in predicting the future. We are interested in making the future. Therefore, we start with today's decisions, because the future is being made today. We would rather focus on who is making those decisions, how they are being made, and how we can provide management services to help make decisions that will allow the future of Kansas City to be as good as the position the city finds itself in today. Here again, management integrates both the present and the future. A good manager, in any situation, realizes that he must consider both the present and future simultaneously in order to run a successful organization. Living in the present and the future is, for a successful management organization, a necessary condition for existence. It is not a choice to be made. Those who are familiar with planning theory will remember that it is broken into planning and implementation and the short range and the long range. There is a lot of discussion that goes on in the profession about whether one should be focusing on long-range planning (and letting the short-range planning go) or whether one should be doing things today, getting things done and forgetting about long-range planning. Well a good manager realizes that both are necessary and that success as a manager, or as an organization, depends on how well both are done. That is to say, it matters how well we make choices to do things today that will not only solve our problems or make us successful today, but that also will allow us to be successful tomorrow. That is the definition of a good manager, and we try to practice that in the department.

Another characteristic of our department is an insistence on adapting and creating technical products, services, and techniques to fit the environment in which they are used. The environment in which local planning and development agencies operate is not static. It is not as predictable as would be suggested by the concepts and techniques that were used in the earlier days, i.e., master plans, comprehensive plans, and so on. Such plans suggest that the environment is static and that we can generate a plan within that environment. The environment in which we operate, however, is very dynamic, open, and continuous. It is highly uncertain. It is incremental. It takes place on a daily basis. It is cumulative; for example, if we build an industrial park on this corner and we put another building on the next corner, then those decisions add up and affect the real estate decisions made the day after tomorrow. It is also what I would call non-rescindable in the short run. If we decide that the airport should be thirty miles from the downtown and we find out five years later that it is not working, then we must live with our non-rescindable decision. We cannot move the airport to another location; it will be affecting development decisions as long as it is operating.

All of this is a lot different from a model cities program or an education program where, if we find that we are teaching people according to a certain theory and that it is not working, we scrap the program, and, except for the costs to the kids and teachers, we can start all over again. We cannot do that in our business.

Another characteristic of our Kansas City environment is that it is driven by the private sector. *We can help the private sector get things done, but the private sector is really making decisions that affect the form, function, physical character, and life within the city.* With very few exceptions, the private sector is calling the tune within American cities. In our business development programs, things are shaped by patterns of investment—that is, investment in buildings, streets, residential units, industrial parks, and all the infrastructure and accompanying facilities that shape the patterns of cities. Therefore, we think the technical products of planning must respond to that kind of environment. They must be applicable to the kind of environment we are operating within. So we have put more of an emphasis on strategic planning, as an important management tool, rather than the more full-blown, detailed management plans that would be part of a comprehensive or master plan.

We are looking for development programs that emphasize what has been called *purposeful incrementalism*. That is, we know that things happen on a day-to-day basis, and we are interested in giving some purpose to those incremental decisions and in guiding them within some kind of general framework. We are also interested in providing programs and services that not only deal with the routine decisions of the city, like the budget cycle, but that also can deal with the ad hoc decisions, the things that are lobbed over the wall at us day after day that we are not ready for. We may not be able to deal with them, but we must deal with them if we are going to be effective as an agency. We are also looking for programs that respond effectively to market forces and that are investment-oriented, especially toward guiding investment by the private sector. We are looking for programs and projects that are action-oriented and that can be implemented, even partially, in the short run because of the incremental nature of the environment in which we operate.

There are two types of innovation that I would like to talk about. The first is the department's management plan. This is a plan, I believe, that makes the management approach operational and provides the means by which the department can make planning professionals more effective in local government. It can continue to keep the department relevant through what I would call natural means. By carrying out this plan, we can assure ourselves that we are always being relevant and tied into what is happening in the city. The other innovation is the products and services that emerge from the department as a result of carrying out this plan. Central business district plans, housing programs, economic development programs, and so on, are good examples. The management plan is intended to help the department become more effective by helping the planning professional provide the highest level of service to the public. By so doing, we can help manage the urban development process that takes place in the city in order to meet the economic, social, and fiscal or financing objectives of the City of Kansas City.

The overriding goal of the department is to help the city as a whole and each subarea of the city work toward performing closer to the potential that exists for the city as a whole and for those subareas. By potential I mean that level of performance that exists in an unfulfilled or unrealized state within the industrial, commercial, and residential areas. It also represents the potential that is unfulfilled among the various groups who make decisions about the city, whether it be the City Council, Planning Commission, local business group, industrial development corporation, or whatever. So our job is to uncover that potential and to try to help people realize it through the programs that we provide.

There is much diversity of development in Kansas City. We are not like a Detroit or St. Louis, but we have all kinds of development activity—from the older central core, commercial, and residential areas to the new development taking place in the central city to areas in what would normally be the suburbs outside the central boundaries. So I think the goal of the department is to move toward trying to capture the potential of Kansas City. The city is now beginning to grow because of its centrality, so this is a very practical goal, and the management plan is a way to achieve it. We have been involved in several major objectives. One is providing services to help people meet the needs they have. We provide a range of management services that includes planning but that may also include finding the necessary resources to carry out the plan, getting the necessary approvals from various departments (for transportation, utilities, and so on); getting the approval of the Planning Commission and City Council; carrying out those plans in some cases by providing product management services and by setting up field offices; and having our own staff help the corporations to carry out the program. We have focused on producing high quality professional products. We try to tailor the products we develop to meet the needs of the particular clients. That means that no two products of the department are really ever alike because we are very sensitive to the needs of the groups in the areas in which we are operating.

An important aspect of these objectives is obtaining results from plans. By locating ourselves properly within City Hall, by making sure that the plans are developed with an action orientation, and by making sure that the client groups themselves are implementation-oriented, we get such results. We start out by thinking implementation and action. From the beginning, we are putting plans together that have an action component in them that is short-range in nature and that can be carried out incrementally. Working within City Hall, we are able to work with the Council and the various departments to make sure that when resources come up and are available that we know that, and we can then begin to get those resources attached to some of the products that we are turning out.

In order to do these things there are a number of policies and programs of the department. They start with developing a competent professional staff. Talent is the basis for the organization. We have a very heavy recruiting program. We have a very strong in-house staff development program. We try to develop planner-managers. We try to use what we call the *star system:* a planner comes in and is assigned a project that is his to manage. Our job is a coaching job to help that person do his best to be successful. We are not competing with these planners because we have been through that route before. We know how to do those things. What we are trying to do now is make the staff successful, and so we are basically in a coaching capacity. We do not get on the field to play very often. We let the planning staff do that.

We try to maintain a very comprehensive mix of products and services. Strategic planning is a major element of our department. We have plans now being prepared for economic development, housing and neighborhood development, and urban design and physical character. These are tied to an Overall Development Program. We have an area planning program that covers 46 planning areas within the city. We carry out a number of project plans where again we work with developers to help them put together zoning packages and to help them design their projects for residential or commercial areas. We try to maintain this very comprehensive mix of services, and we also try to stay at least a half a step ahead of everybody else. In our business, turnaround time is very important, that is, if someone asks to be provided with a service, they would like to have that service right away because they desperately have that need. So what we do is try to anticipate what those needs and requests are going to be and pull those into our work program. At any time, we may be working on products that no one knows we are working on, products we think people are going to ask for in six months or a year—when they do, we can give them results very quickly.

I would like to close by describing a couple of the programs that we are involved in at the moment that are most interesting. The Alternate Futures Program for Kansas City is a program that we carried out with the Brookings Institution. It was the first attempt since 1947 to try to take a comprehensive look at Kansas City and its region. What we did was put together analyses for economic development, housing, population, energy, and education; working with a seminar of about 100 people, we involved a larger group of 1,500 people through questionnaires. The results of this program were published, and a resolution was passed indicating that the functions that were identified were very important to Kansas City and ought to be pursued. A program is now being put together with the private sector to go into more detail and to carry out programs in each of those functional categories.

The Development Program is one we have been working on with a budget and systems group within the city. People have been asking me when we are going to do another master plan for the city. My answer is never, or at least not as long as I am Director of City Development. What I would like to do instead is to develop a Citywide Development Program and to do that on an annual basis, that is, similar to the one mandated by the Charter in Atlanta, but we do not have any mandate like that in Kansas City. I believe, again from a management point of view, that the Council and all other people who are involved in investment decisions ought to be able to look at the state of the city once a year and be able to make decisions on an incremental and annual basis on where they ought to be spending their resources that year in order to meet our development objectives. We are proposing this, and we have just started a program to put together a multi-year proposal that would get this system going in the next two to four years. It will be a very large task. We have been building a capability over the past couple of years, and we are now into this program.

We have a general development plan for the Troost neighborhood, which is a good example of the kind of work we do with client groups for our client-centered activity. Here we worked with an association that represented most of the owners of property and business firms in the older, decaying commercial area of Kansas City. We helped them get funds to develop and plan. We worked with them to get consultants whom we managed in carrying out this plan. We helped them get all of the approvals they needed from city departments and from the City Council, and we now have the project manager who is in office in the Troost midtown area helping them. He is putting together a series of public improvement proposals, a streetscape program, and plans for the purchase of some areas for parking lots because that was their major priority. They felt that they needed to locate parking more strategically to their businesses. About half of that manager's time is spent putting together private development packages, looking at some of the key buildings in that area, trying to find tenants, trying to get resources from the Small Business Association and other loan funds that are available through the city, and actually putting development packages together to get tenants into this area to fulfill the function we have identified for it.

The central industrial district is another development plan for the old stockyards area. Here we worked with CIDA (Central Industrial District Association) which is made up of the landowners and firms in that area. We were able to get EDA technical assistance funds to develop the plan working with the CIDA. We have this plan approved now, and we are in the process of finding funds to put in about $6 million of public systems, arterial roads, and flood prevention systems in order to make this area more attractive.

The final program that we are embarking on is an Urban Economic Development Council. Kansas City was one of ten cities selected under a new demonstration program put together by the Departments of Labor, Housing and Urban Development, and Commerce. What we are doing here is creating a new institutional arrangement for economic development in the City of Kansas City tied directly to a new technical advisory committee, which the City Development Department chairs. This is a new effort to try to bring together the very diverse interests in economic development in the city and to focus on the special impact area—the older area—in order to put together a multi-year program as well as to use the resources that we get from Housing and Urban

Development, Department of Labor, and EDA in more effective ways.

QUESTION. I would think it is fair to say that your approach to planning in Kansas City is in favor of development. Development potential in many cases is the object of the planner. Yet Bonner believes that the object of planning is sometimes to discourage development. Do you have those kinds of pressures in Kansas City? Are there some groups of people who do not want this development and would like to see Kansas City stay as is?

RESPONSE: Overall, Kansas City is not faced with growth pressures. Our basic economy grows at about 2% a year, so we are not a rapidly growing economy; we are about at the national average. Professor Wilbur Thompson came to talk to us about our economy. He said that if he could plan an economy he would plan it like Kansas City's because it is very diverse; it is slow growing; and it allows us to plan for change without being overwhelmed by that change, either for industrial employment activity or for the residential population that comes with it.

We have just put together what we call a City Wide Viewpoint which tries to define what the central tendencies will be for the city for the next 25 years. It is clear that we are getting the best of both worlds. We are not losing the industrial growth that northeastern cities are losing, and we are holding on to most of our manufacturing jobs. On the other hand, we are moving more toward a Sunbelt economy—more toward distribution and services—but we are not getting all the population that comes with it. We are getting the jobs and the investments, but we are not getting enormous numbers of people. The growth of population is very slow. So we do not have those kinds of pressures.

In any particular situation, there will be some community groups who do not want another retail center in the neighborhood, or who do not want another industrial park next to them, or who do not want the landfill site over the hill. Those are the kinds of things we have to work out with groups on a day-to-day basis. We just finished a plan for a suburban area of Kansas City as one of our 46 area plans. What we did was to suggest that we take out a commercial center because the residents convinced us that it was not the kind of thing they needed. So, we said, "OK, we'll take it out of the plan, we'll restudy it in three years, because by that time the whole commercial development industry may change or you people may want the commercial center by that time." We will respond to the need to protect the residential environment by not recommending such guides for Council decisions. We do not have, however, the kinds of enormous problems that they have in growth areas like California, Florida, or Texas.

Q: It sounds as though you take very much of a public entrepreneurial role in the City Development Department in Kansas City. Who determines the projects that you are going to undertake? Who determines, for example, the overriding goal that you mentioned of helping the city reach its potential? Do you choose that yourself and do you then take it to the Mayor?

R: We basically determine that overriding goal by ourselves. That is probably a very positive characteristic of the department. When the City Development Department was created by ordinance, back in 1968, we were required to provide services to the Planning Commission. The Planning Commission in Kansas City has certain responsibilities. It is responsible for the Comprehensive Plan of the city. It is responsible for reviewing all zoning cases and subdivisions before they go to the Council. But there are a lot of programs, like the Troost Midtown or Economic Development Council, that we just set up. The central business district programs, which are not required to go to the Plan Commission, allow us to be very free to decide what to do. There are certain requirements that we must fulfill—the zoning reviews, for example—but outside of that, we can determine our own future.

Now, having said that, I am also convinced that the way for us to be more effective is to go toward an Annual Development Program, in which we can be of more service to the Council by helping them to see what trends are taking place in the city, where the emphasis ought to be for development, and how the department ought to be carrying out its own programs. I think that will help the department in the long run to become much more institutionalized within city government and remain effective for a longer period of time. I am glad we are doing the things that we are doing, but I really am leaning toward becoming more a part of that Annual Budget and Development Process.

Q: On a visit to Kansas City earlier this year we stayed at the Royal Crown Center, and, as nice as that whole development is, I understand that some of the downtown people were unhappy because this development is some distance from the downtown area. They believe that some of the development should have taken place downtown. Has this problem been resolved now?

R: No, it has not been resolved, and I do not know that it ever will be. You are right, there is a central business district, and, of course, the Royal Crown Center has an enormous amount of office and commercial space in it. At the moment there is competition for a Hyatt Regency Hotel. There is a site that is available in the Crown Center Development Plan, but our development plan for downtown also suggests that there need to be at least 1,400 to 1,800 more rooms in order to make the Convention Center more effective. We have a gentlemen's agreement, at the moment, with Hallmark in terms of carrying out the necessary studies in order to develop a hotel package. But clearly that competition is going to remain, and I do not know what the city can do about it. It is basically a private sector competition.

Q: We noted that the Muehlebach Hotel was being refurbished at that time, and they promised some more hotels down there. You also have some hard-core porno shops, x-rated movies, and whatnot down there. Is it a part of the planning concept to phase them out or is there any effort to get rid of these types of activities?

R: I think we are probably divided on that. There are some people for it and some against. I would say there is a Downtown Development Program, and our emphasis there is trying to take advantage of the $30 million or so that has been invested in the Convention Center. The Radisson Muehlebach Hotel has invested about $7 to $8 million in renovation. They have done a tremendous job in that. We are now programming about $2 million of Community Development funds in the downtown area. We will do things to enhance the downtown pedestrian in the human-scale environment through skywalks, pedestrian ways, transit ways, and linking up some of the major activity centers downtown. At the same time we are trying to attract private development.

There are two major projects downtown. One is the downtown hotel project, and we have been actively working with about five hotel chains, including Hyatt-Regency, to try to put together a package on the site just across from the Convention Center. We are also looking at a major commercial renovation project that would tie together the major anchors downtown—i.e., Macy's and Jones, the two department stores—with another kind of galleria of commercial space. There is no question, though, that we are competing with Crown Center in that respect. I think that both of them can exist. I think what will happen is that they will both develop at a much slower rate than they would have otherwise, because we can only absorb so much office and commercial space.

Q: You talked about developing some sort of increase in the

residential use of the older part of the central business district. Why do you think people would want to live in the old downtown? What sort of economic class are you going to attract?

R: The market for downtown residential land is basically households without children, so we will have a lot of elderly people in the downtown and in some of the areas to the east and west of the downtown core. Within the central business district, we expect to have some younger singles and families without children, people who are very interested in some of those residences downtown. That is the market we are looking at. We are not looking at a market for a family with children, because the school situation is not good there. However, we have had inquiries from a couple of developers looking at the potential for rehabilitating some housing close to the Convention Center.

Q: What development tools do you have to use in Kansas City in your activities? Are you actively engaged in putting together legislation that might be useful for cities in Missouri such as Kansas City? Is that an active part of what you do?

R: The main pieces of legislation that we use for development in Kansas City are Act 353 and the Planned Industrial Expansion Act. The 353 Act allows a developer to use eminent domain powers and tax abatement (on a 25-year tax abatement schedule) to help redevelop blighted areas. Crown Center was developed under the 353 Act. The Planned Industrial Expansion Act is for industrial projects. It is used to finance industrial buildings, so it is really a bond program. The Industrial Commission can sell revenue bonds to construct facilities, and they use the credit and strength of the individual industry that is going to occupy that building as the basis for being able to negotiate the sale of the bonds. Those are the two main vehicles we have.

The Development Department does look at legislation, and

we are looking at legislation programs for housing and economic development. I proposed to the Council this summer that we look for new ways to finance major redevelopment efforts because we can begin to see there is a decline in Community Development Revenue-Sharing, and that money is not being used the way it was used when it was categorical. So we do not have any major redevelopment projects underway with the $17 million a year that we are using now. I think that we need to develop a resource so we will be able to fund redevelopment activity. We suggested that they look at tax increment financing. Kansas has just passed such a bill, but it is not a very good bill. I do not think it can work. I do not think we will be able to use it in its present form. Legislation is very much a part of our interest in making the city competitive for industrial development, housing, and so on. We do a lot of work with the subdivision regulations and zoning ordinances for builders to try to help them market land and projects in Kansas City.

[margin note: Controversial]

Q: What is the state of the residential housing stock, and what efforts are being made to revive it, as I understand there are some sections which are below code?

R: We have done a lot of work in that area. One of the first programs the Development Department got involved in was trying to put together an experimental program for housing conservation. The Council, under General Revenue-Sharing, back in 1973 set aside money for a Conservation Demonstration Program, and they asked us to design it. We worked with them for about four months putting together three demonstration areas. We have put about $800,000 into each of those areas. We set up a Management Committee made up of (1) the Council members from the district, (2) various city departments that had involvement in conservation and public facilities development, and (3) residents; about half of the people on that committee were residents who represented that particular community. They worked with the staff, and they decided how the money ought to be spent and for what priority items. The program is built around systematic code enforcement. It is also built around loans and grants. We used the Federal 312 Loan Program for a while, but we will be using a new rehabilitation program that has just been set up through the city. It is a $3 million program to support these areas. In one of the areas we created a Neighborhood Housing Services Program funded through the department. We have dealt with the Real Estate Research Corporation to try to develop a housing typology—that is, to try to classify areas based on their market, conditions of the housing, population mix, location, and so on. We have begun to tailor-make programs for each of these areas. For example, in the suburban parts of Kansas City, we have a surveillance of residential areas but not too much in the way of code enforcement. In the inner-city areas, we have code enforcement, systematic public improvements, and, perhaps, help for community groups organizing themselves so they can begin to be more effective in managing change within that community. I think in many of the areas it is going well.

The areas we are having problems with are in what we call the "Type Two Areas," which are the low and moderate income areas in which there is a lot of abandonment. There is a lot of Section 235 Housing in those areas because they were flooded with that kind of housing, and those are having a detrimental effect on the neighborhood. There are families who are low income and cannot afford even the low interest loans that we are providing in the program. We intend to do more with that, but it is a very difficult situation. There is just never enough money to do what we want to do.

Q: What do you think we can anticipate in the way of federal policy with a change in the federal government? What kinds of changes would you advocate in terms of federal policies?

R: Our job is to make the future, not predict it. I am not going to try to predict what future federal policy will be. But

I certainly think that there ought to be more emphasis in the federal government on economic development. We have been spending a lot of time on economic development lately, and we realize, for example, that a lot of the older employment areas in cities like Kansas City are declining. We find that when we look at tools to help us in conserving, rehabilitating, and renovating these areas, we do not have the array of tools that we have for housing and conservation. For example, we have the Section 312 Loan Program to help aging neighborhoods in terms of low-interest loans. We do not have that kind of program for industrial or economic development. We do not have any systematic way to go about serving industrial areas the way we do residential areas. So I would say we need these kinds of packages.

Q: What kinds of restraints and constraints, if any, have you had?

R: I guess the main restraint comes from the value system of Kansas City. That is, it has long been believed—and I think it is true for the most part—that the private sector does everything and they always take the leadership. So the constraint in city government in Kansas City is that we generally have to wait for the private sector to take the lead. Now, being a public entrepreneur, I do not believe in that completely, but I understand the situation; on the other hand, we are generating a lot of interest in the private sector. We are doing it in a lot of different ways. That itself is a constraint. We do not operate the way they do in Boston, Chicago, Cleveland, or Detroit, where they have really serious problems and the public cannot afford to sit around and wait for the private sector to act. They have to jump in and do something radical. We do not have that in Kansas City, and, therefore, the private model still exists—that is, the private sector is out front and the public sector is there to help. We take that posture, and that is a constraint on what we do—but not too much of a constraint.

Q: Are you ever told not to mess with something by the Mayor or the members of the Council?

R: Sure, but we generally do get into everything. That is just the nature of the game we are in. The City Development Department is seen as the agency in City Hall where people come when they have some problem with development. People are always coming to our doors, and we can decide how to handle their situations.

There is a tradition in city government in Kansas City. When I came they referred to major departments as little fiefdoms, as kingdoms all their own. For example, the Parks Department is a fiefdom because it has a separate makeup. The Park Board is appointed by the Mayor, but then the Park Board appoints a Director, whereas the City Manager appoints all the other directors. That Director is separate, and that was done because the Prendergast Machine felt that it had a tremendous park system that might be taken over and used for different purposes, so they made it a very separate part of city government. It is very difficult for us because there are two kinds of things going on in city government. One is the functional activity—transportation, public works, parks, and so on—but then there is the integrated function, which is where we are—that is, we are trying to integrate, in any given area, all of these elements, plus the private sector development, to put together a package that works. That is a problem within the city. We are gradually overcoming that, because I think we are able now to show people that we have the capacity to work with them at their level. For many years the Development Department was not respected because it did not have the staff, could not compete, and could not run with the other departments that had good professionals.

Q: Your department deals with the zoning ordinance. Does the Building Inspector's Office also deal with the implementation of the zoning ordinance? Do you ever have conflicts in your interpretations of how it should be implemented?

R: The Building Inspector is in the Public Works Department and issues building permits. We really have not had many conflicts. What they do is refer to us. Many of our zoning cases go through with conditions, that is, they are Planned Unit Developments, whether they are industrial, commercial, or residential, and the Plan Commission and Council put numerous conditions, say for landscaping, setbacks, or parking, on those developments. The Public Works Department has been burned a couple times by allowing a permit to go ahead and then finding that the residents are up in arms because something was happening that they understood was not going to happen. Now we have a different arrangement with the Public Works Department; they refer everything to us and we review it for the conditions. Then we tell them if it is all right to go ahead and issue the permit from the planning point of view, and they will look at it from the building construction viewpoint.

Q: You make everything sound good in Kansas City! So what are Kansas City's major problems?

R: The major problem the city has is being able to provide a continuous flow of funds for investment in development. That is, we have in Missouri the requirement that all bonds must be passed by referendum, and that requires a two-thirds vote of the population, so we have not passed very many bond issues in the last couple of years. That is causing severe problems with respect to just maintaining the physical plant that we have without going into development or expanding the city's capacities in any way.

In Michigan, the state legislature passed a law that allowed the City of Detroit to spend a certain percentage of its overall assessed valuation with capital funding that amounted to about $27 to $28 million every year. They are able to plan three-, four-, and five-year programs incrementally and get things done. We do not do that in Kansas City. We must go with a big bond package, and the way that bond package is put together is not always the most logical way to do it. In the last couple of years we have had several bond programs fail in referenda. They did get over 55% of the votes and had a majority, but they did not get the two-thirds vote required by that law. There has been a move afoot to get the legislature to change that, but it has not been successful to date.

Kansas City really does not have too many problems. I think there is certainly an educational problem. Again looking at it from the development point of view, the city annexed too many school districts. Within Kansas City, there are thirteen schools districts that we have to deal with. The largest and most urban is the Kansas City School District, and that is a problem district that is affecting our ability to attract middle-income families into the older areas. The Country Club Plaza Area within this school district, for example, is a very well-planned area with substantial homes that are fifty to sixty years old. We know that in the next five to ten years there is going to be a lot of turnover there because it has the largest percentage of elderly population of any of the census tracts in the city. The question is, what market is going to replace the elderly market that is there? Will it depend greatly on the Kansas City School District and our ability to solve the school problems? That is a major problem that a lot of people have been working on, but it is not solved.

Q: To what extent do Kansas City, Missouri, and Kansas City, Kansas, work jointly on long-range economic and social development projects?

R: We do not do any joint planning with them! It would be more fair to say we are in competition with them. Kansas City, Kansas is, of course, much smaller in terms of population and area. Their economy is much different than ours. We really have not had the occasion in all the time I have been in the city to do much work with them.

Q: What are some of the strategies that you are taking for the downtown retail area?

R: One of the other techniques we are using besides Act 353 and Planned Industrial Expansion is the Special Benefit District. The state had a Special Benefit District Law, but it was very cumbersome and very poorly drafted. It required that the majority of the owners of property had to petition to have the benefit district set up. It also required then that a majority of the residents had to vote for it, and it was very unclear as to who was an owner and who was a resident and how we were to get them together. We changed that bill and helped draft a new bill and also supported it in the state legislature. This allowed us to have any one person in the district petition the Council to set up a district. Then the Council would hold public hearings, and they would either approve or deny that petition.

We have been working with the downtown retail people for the last six months, helping them put together a special benefit district package. The amount of money that would become available to the downtown retail core would be about $800,000 a year under full assessment (or, realistically, about $400,000 because we do not think the retail people and owners are going to accept the full levy that is possible under the bill). For about $400,000, we are looking at a management program for marketing the downtown area using some of the funds to add services above what the city is providing, for example, in security or extra cleanup.

We have a streetscape program that we are putting together with Community Development Funds. We think that some of that money could be used to pay for a joint funding of downtown programs. They are looking for a downtown development manager to create events and make things happen in the downtown area on a weekly basis. We are trying to get the retail group to organize themselves into a special benefit district and maybe even organize themselves as a corporation, using that money as an annual incremental flow to market the downtown area. Maybe they can put a program together for acquiring certain sites using Act 353 in certain places and rehabilitating certain buildings.

Q: In terms of that approach it seems to be, as you said, an incremental approach, and that it is hard to deal with the big draw from outlying shopping centers. Do you think an approach with more impact, let's say getting an outside developer with another department store, would offer more strength to compete with shopping centers?

R: The Galleria project is that kind of project. What we are saying there is that the two blocks are strategically located between Macy's and Jones, which are our two department store anchors downtown. They are right across the street from City Center Square, which is a major new office building with about 700,000 sq. ft. of office space, and a retail center attached to it across the street from the new Mercantile Bank Building built last year. What we are trying to do there is work with developers. We have been talking to some of the downtown realtors about it and, in fact, we have people interested in trying to get a group of financiers together to form a corporation to look at the possibility of coming in, probably under Act 353, to acquire both of those blocks and put that project together. They are very much interested in the concept of the Galleria, and they want to see if they can form a group that will have the right kind of backing. It would fit in with the retail plan that we have with the special benefit district. But you are right in that we have to have both of these going on. In some cases, massive investment is needed to turn something around. This would be one of those cases.

Q: Kansas City seems to be blessed with more than the usual amount of entrepreneurs who want to invest in the downtown and in new development. Am I right in believing that?

R: No!

Q: Then how do you encourage them? Only by the use of these planning techniques?

R: Some of the major development projects have been like the Hallmark Project, Crown Center Development Project, that was financed by the Hallmark family, but we have not had that kind of investment in the downtown area. It has been very slow in coming, and we spend a lot of our time trying to tap investors, to find people with the necessary equity to venture a project like this. We have been working on a hotel site now for over six months. It is a bad time to try to put a hotel package together because one of the riskiest adventures is to finance a hotel. We just do not have that kind of equity around. The Galleria project would be the same way. It is not true that there are a lot of people out there wanting to invest in the city. We have to struggle and fight for every investment we get.

Q: There seems to be an indication that the occupancy rate in downtown hotels is quite low already, in the 40% range, and that strikes me as not a particularly viable market for a large hotel complex unless it is going to bring its own clientele. There does not seem to be a demand for hotel space downtown. Why pursue it?

R: There is a demand for first-class hotel space. Remember that the 40% occupancy rate is for the second-class and third-class hotel facilities downtown now. We have talked with the Convention Service Bureau often about what kinds of conventions we are attracting and why certain conventions do not come to Kansas City. Many times we are missing a whole level of conventions because we do not have the first-class hotel rooms that we need. They just do not have enough of that kind of space. The Muehlebach Hotel renovation will help, but we still need 1,400 to 1,800 additional rooms. People are asking for rooms close to the Convention Center, that is, they want to be able to walk to the Center from their hotel room.

So the demand is there, but the real problem now is the financing. The financing picture is not good for that kind of development. The Hyatt-Regency Hotel, for example, is a $55 to $60 million investment, which requires $10 to $12 million in equity. Now there just are not many people willing to put that kind of money down to go into that kind of venture. There is also a cost push, that is, in order to build that facility, it is necessary to charge $55 to $60 a night. It is going to be tough for any hotel management firm to be able to keep up the 70% occupancy rate that is needed. So it is very difficult to build them today, but we believe it is essential for Kansas City.

QUESTIONER'S RESPONSE: It would seem so!

Chapter 6

MILWAUKEE: PLANNING FOR FISCAL BALANCE

WILLIAM RYAN DREW
Commissioner of City Development
Milwaukee, Wisconsin

I am pleased to have the opportunity to talk about planning and programming in the City of Milwaukee. We have found some new ways, I believe, to integrate our comprehensive planning work with our program planning and operations—and we are very excited about it.

Milwaukee, like most central cities, has been losing population steadily for the past fifteen years. But we have had a modest increase in our number of households. Thus, we are losing families with children but gaining adult households.

Unlike most older central cities, Milwaukee has a lot of vacant land—close to 8,000 acres. Most of this land, annexed at the periphery of the city in the early 1960s, is not fully served with sewers and streets but is ripe for suburban-type residential and industrial development. We also have cleared and underused land—appropriate for residential and industrial reuse—near downtown. So we have new development and redevelopment possibilities—but not enough expected growth to justify strong emphasis on both.

Like other central cities, Milwaukee has only a small amount of money to spend on new development and redevelopment activities and only limited influence on private market development and redevelopment. But we know that our money and our influence must be consistently used if we are to reach our end goals.

To accomplish this, we are using a dynamic comprehensive planning process linked directly to our program development and operation activities. We have the beginnings of a system that will allow us to use a unified approach for programs in housing, commercial and industrial development, public improvements, code enforcement, transportation, and downtown redevelopment. When the city government has its own planning and programming in order, it should allow us to intervene effectively in the private market.

Our comprehensive planning and programming process in Milwaukee involves (1) a computerized planning information system, (2) new analytic techniques and tools to measure and monitor the condition of our city, (3) a policy development procedure that encourages consistency between short- and long-range goals, and (4) a program development posture geared to move us incrementally toward our long-term goals.

Before I explain these elements, let me emphasize three key concepts that underlie our comprehensive planning efforts.

First, we have not finished our comprehensive plan—and we never will. We believe that our city changes too rapidly to be tied to a printed and bound copy of a "Master Plan" as we have known it in the past. We visualize, instead, a set of maps showing existing conditions and potential development, which can be updated as circumstances change, and a text of policy statements based on an analysis of the current state of the city. The key concept is "dynamic" not static.

Second, because we will never have a finished "Master Plan" document, every element and every activity in our comprehensive planning process must have immediate utility to program planners and policy makers. We cannot afford to take five years to complete a document for reference; we need information now to guide our development and redevelopment activities. The concept is to generate information that is practical and immediately useful.

Third, all of our operational programs must be designed to further our long-range development goals. All of our departmental staff, in fact all of our city departments, must share the same mental picture today of what the city might look like in twenty years. While our goals are dynamic, too, we cannot have our long-range planners working in one direction and our building inspectors and urban renewal planners working in another. Our short-term activities must be consistent with our long-term objectives.

The first step in our comprehensive planning activities was the creation of a computerized planning information system. We started with information describing the physical state of Milwaukee. This included land use, number of parcels, and structures. We received information from the tax assessor on types, characteristics, ages, and values of structures. The building inspector provided data on housing code violations, demolitions, new construction, and changes in use. The Department of City Development added land use and zoning codes and parcel area. Our initial emphasis was on residential land use, since that is the largest use in the city, but we are now adding data on commercial and industrial structures and uses.

We are very pleased with the promise of our planning information system, but I do not want you to think that its development was a simple task. Just developing a uniform addressing system and getting other departments to record data that we need and they do not is a monumental job. For example, we need complete and accurate information on tax-exempt property, but that is not a high priority item for the tax assessor. But rather than dwell on the problems, let me cite a couple of real successes. In the process of building our demographic and socioeconomic files in the planning information system, we have been successful in obtaining unemployment data from the Wisconsin Department of Industry, Labor and Human Relations, and annual income data in sub-areas of the city from the Wisconsin Department of Revenue. With cooperation like this, a truly comprehensive planning information system is possible.

As we continue to build and refine our physical, demographic, and socioeconomic data files and thereby improve our planning information system, we also continue to develop techniques and tools to massage the data. We have brought together an interdisciplinary team for these activities—a geographer, an economist, a transportation planner, an environmentalist, and a social scientist.

The first hard product of our data analysis has been a relative residential status index. We analyzed ten variables (dealing with such things as home values or rent levels, percentage of owner occupancy, length of residency, and overcrowding) and developed six categories of relative residential status. ("Relative" is an important word because determining neighborhood condition is always subjective—and a neighborhood can only be described in relation to other neighborhoods in the same city.)

Milwaukee is now mapped in terms of its relative residential status into areas which are: most stable; less stable; early decline; visible decline; clear decline; and established decline. Our most stable neighborhoods exhibit rising prices, readily available conventional mortgages, normal turnover, and high levels of maintenance and repair. The subsequent stages demonstrate progressive degrees of deviation from "normal" real estate market conditions. Typical "abnormal" indicators are lack of mortgage availability (or use of government-insured loans only), stable or declining values, deferred or neglected maintenance, subdivision of large units into a number of smaller units, high vacancy rates, a high percentage of renters rather than owner occupants, vandalism, and abandonment of buildings.

A map depicting our relative residential status has already been useful in selecting urban renewal and neighborhood preservation areas and in helping us decide where to spend our Community Development Funds. A second analytic product is due shortly and will add another dimension to our decision-making base. This product—a demand/need index—will add a socioeconomic overlay to our physical map. The demand/need index looks at the city's population in terms of life cycle stages (age, number of children, household size, mobility rates, and so on) and socioeconomic status (basically occupation and income).

With the two indices mapped at the block group level, we will be able to delineate functional planning districts to help us plan our housing and economic development activities. For example, we can pinpoint (1) areas with declining values but moderate-income households who can afford to maintain their structures; or (2) small pockets of beginning decline surrounded by stable neighborhoods—because the beginning decline neighborhood has a high percentage of low-income elderly owner occupants; or (3) stable areas in need of additional commercial services. The programming implications from this type of planning analysis are, I believe, immense. Not only will the indices enable us to identify areas suited to various types of programs, but they also will allow us to monitor the effectiveness of these programs over time.

Simultaneously with the development of our planning information system and our analytic condition and monitoring measures, we began our policy definition work. *We decided that fiscal balance would be our overall goal and the cornerstone of our comprehensive planning and programming efforts.* We believe that the principle underlying overall planning for any jurisdiction should be fiscal balance. Over time, the trend should be toward having revenues equal or exceed costs. That is a deceptively simple statement, as you know. I am not suggesting that central cities can be self-supporting. They must be net beneficiaries of state and federal revenue-sharing. But they must be able to function from year to year with the revenue-sharing that is available—which will always be too little. This means that we have to be very conscious of long-term trends and of fiscal relationships among our community activities. As cities age, they tend to house relatively poorer households. As their capital improvements age, replacement and operating costs increase. This usually means that redevelopment to higher density uses is necessary to increase the tax base enough to support capital, personal, and property service costs. Redevelopment raises policy questions as well as market questions. How much redevelopment could actually occur, and how much of what kind do we want if we can get it?

As a method of bringing together the policy and fiscal concerns of city development, we are analyzing what we call the functions of the City of Milwaukee. By "functions" we mean the essential activities of the central city. As we see them, these cover three basic types of activities:

1. The first is basic economic activities that contribute to the overall vitality and revenues of the city. These include attracting and keeping employers; providing centralized downtown facilities for offices, retailing, hotels, conventions, and cultural events; and maintaining attractive environments for the city's residents. Most of these functions provide more revenue than they cost.

2. The second is basic services to the resident population. These functions include providing educational, governmental, medical, cultural, and recreational services; and upgrading low-income and otherwise disadvantaged households economically, socially, and culturally. Most of these functions are revenue absorbers, though some have mixed fiscal effects.

3. The third is the maintenance or "housekeeping" activities that sustain the other basic functions. These include transportation, police and fire protection, waste disposal, and public works. All of these activities absorb revenues, and the costs rise steadily.

The main reasons for looking at city operations in a functional way are to be aware of the fiscal implications of ongoing activities and to develop programs that will accelerate or enhance revenue-generating programs and balance out the steadily rising costs of the necessary service functions. We have just completed detailed revenue/cost analyses for the City of Milwaukee that show that our residential land uses—

citywide—do not pay their own way. That is not a surprise, but the degree of fiscal drain by families with children in public schools was.

With 1.0 representing balance of revenue and costs, our residential land use is at 0.7. With commercial at 2.0 and industrial at a whopping 3.2, we looked more closely at the residential category. Adult households (without children) pay 1.5; households with children not attending public schools pay 1.7; but families with children in public schools pay only 0.3 of their costs.

Obviously, one cannot plan a city that emphasizes industrial, commercial, and adult-only households (in that order) and that avoids families with children in public schools. But this work does give us a basis for careful planning for our vacant land. We must encourage commercial and industrial development to cover the costs of new residential construction; we must provide incentives to redevelopers interested in high-density development near downtown or near our universities because they will attract adult households with middle and upper incomes who will be net revenue producers. We must do both of these things at a large enough scale to be able to afford to encourage family-oriented housing as well.

This is a superficial sketch of the fiscal balance approach to our comprehensive planning efforts, but I think it gives a sense of our orientation. Revenue/cost work is time-consuming and not very precise even when done carefully. But it does not have to be exact to provide general guidance for policy-making. We believe that awareness of long-term trends and their implications is very important and that aggregate effects are too often neglected when attention is focused solely on day-to-day programmatic issues.

What are we learning from these planning efforts and how are we applying this information in our program operations? We do not have our functional planning areas neatly defined and categorized with a battery of remedial programs yet, but I can give some examples of our incremental programming efforts.

We know that our urban renewal funds are limited, as is the immediate potential for massive redevelopment in the center of the city. We cannot solve the problems of our "established decline" neighborhoods with all of our community development and capital improvement money—let alone the portion available for renewal. But spot renewal in "early or visible decline" neighborhoods can be a stabilizing influence, and removal of the worst blight near downtown (in conjunction with a residential land banking program) can offer the possibility of residential development for adult households in the future. Thus, we can capitalize on the trend of growing numbers of adult households and add revenue-generating households at the same time.

We know that we must walk a fine line between new development of our outlying vacant land and redevelopment of our vacant and underdeveloped central-city land. We have an industrial land banking program in our annexed areas, and we are in the initial stages of a massive industrial redevelopment project in a declining industrial area adjacent to our central business district.

On the residential side of the development versus redevelopment question, we know that we should plan for higher density development near the center of the city and discourage high-density residential projects in our outlying areas. If Milwaukee has the potential to compete with our surrounding suburbs for single-family owner occupants, we must do it on the vacant land on the northwest side while using our existing capital infrastructure near downtown for adult household high-density development.

We also know that certain kinds of industries can benefit from a central-city location (in terms of labor force and transportation). Others are more suited to an outlying suburban-type setting. We have something to offer both.

We know that it is important to keep the middle income families we now have—and we know that many of them live in stable, less stable, and early decline neighborhoods. In these areas, we must encourage owner occupancy, good municipal services, and investment in home maintenance. We must also maintain a high level of commercial services to support the residential character of the neighborhoods. We have a battery of neighborhood preservation programs geared to keeping our "good" neighborhoods from declining, and we are working with owner occupants in "declining" neighborhoods to reverse the downward trend.

Our neighborhood preservation program is based on two rather uncomplicated concepts. One is that housing preservation is to be approached as an economic development concern, not as a housing problem per se. Therefore, all housing preservation activities are designed to achieve economic objectives. Neighborhood preservation must take into account not only the residential community, but also the commercial service system which relates to it. Both of these concepts stem from an awareness of the neighborhood as a micro-economic unit which must be considered technically, as well as emotionally, in order to understand the problems and to solve them.

I wish I could bring out our blueprint for all of our activities. Two things prevent this. First, we do not know all the pieces of the puzzle—yet! Second, for those we have defined—we have not written it all down. My staff gives me a choice: we can write down what we are doing or we can keep doing it. Needless to say, I always choose the latter. But soon we are going to be forced to get it all down on paper. We have combined every federal grant we can find—HUD 701, EDA 302, Community Development—and we are applying for an innovative projects program grant to further our planning and programming efforts. And we all know how much on-paper documentation the federal government requires! In the meantime, I can only offer somewhat dated papers—and my willingness to answer any questions that I can.

QUESTION: Certainly one of the major issues of public policy here in Milwaukee has to do with the transportation system, particularly the freeway completion. The Southeastern Wisconsin Regional Planning Commission has a land-use and transportation plan that still projects completion of many more freeways than currently exist. Have you used this process or do you anticipate using the process that you have outlined to address the broader issue of transportation—not just freeway planning but transportation planning—so that the department might make some major recommendations to the Mayor and the Common Council in terms of transportation planning that could possibly conflict with the transportation plans of the regional agency?

RESPONSE: We are involved in transportation planning only in a very peripheral manner. The county obviously is the agency that has purchased the Transport Company, and it is the County Commission that received the gift from the city to plan the expressways. A number of years ago the city took itself out of expressway planning and construction and that became a county function along with the mass transit system. We do get involved in commenting to the Council, to the Mayor, to SEWRPC, to the county, and to other agencies since we feel that we have some responsibility to protect the best interests of the citizens of the City of Milwaukee. We may see what we think is in the best interests of the city a little bit differently than some other agency. As a matter of fact I am pretty sure we would. We only have one individual on the staff who is thoroughly trained in transportation, so we are somewhat limited. We cannot compete with other agencies, but we do comment on their plans and efforts.

I think that maybe you see a conflict between what we would be saying and what the regional agency would be saying or what the county transportation planners may be saying on specific issues. I think the answer to that is yes. I think you will see that in the not too distant future. I do not think that all the expressways that have been planned are going to be built. I say that not from a planning sense at all. There is not enough money to build them. I do not think the Park Freeway West will ever be built because I do not think there is enough money in the county to build it. Outside of the other ramifications, that is just a plain, pure, practical political answer, and that well may be true of some others also.

Q: Who decides what and where money will be spent, and how do you avoid possible conflicts of interest? Do you feel that these people should be under a strong code of ethics?

R: The Common Council decides where the money will be spent. Major activities are conducted by, for instance, the Redevelopment Authority. Any plan that is passed by the Redevelopment Authority has to be approved by the Common Council. Money for activities has to be approved by the Council. The grants have to be accepted, and we have to have permission from the Council to apply for these grants. I think they are under a strong code of ethics.

Q: We have heard a lot in this series about the role of citizen participation. I was wondering how you view that question and how you handle it in the department?

R: It depends on what part of the department we are dealing with, but, as an example, let us take the Redevelopment Authority and the money that comes in from the Community Development Block Grants. To get from stage one where the city receives the money to the point where we spend it on an urban renewal project requires action by the Community Development Agency, normally in cooperation with the Department of City Development. We have a series of public hearings around the community, soliciting ideas on what should be done with the money. We present certain information to the citizens at that time, indicating generally what our thoughts are on where the money might be spent, and we get feedback from that. There is a citizens' advisory board that reacts to the suggestions and makes further suggestions or suggests deletions in programs. The City Council then holds a series of hearings on the plan through its committee system, and it is finally voted on.

Then the Redevelopment Authority goes out into a neighborhood and holds a public hearing. We distribute information on the meeting in a very wide area. Many more people than those immediately affected by the proposed plan are involved. Then there is a formal hearing. At all of these meetings, we have the potential for citizen participation and comment. The Redevelopment Authority then acts on a redevelopment plan and sends that plan to the City Council, which holds another hearing, and the plan is finally voted on. In the first and second years of Community Development funding, we had substantial changes made in the plan: some locations or ideas were dropped and in other cases the configuration of the proposed area was changed based on comments that came in from the citizens. That is no threat as far as we are concerned. We are happy to have people come and tell us what they think.

Q: A number of years ago, when Milwaukee instituted an industrial land bank, it represented a fairly unique endeavor for cities in the United States. It is my understanding that the city, with the passage of the recent state law that exempts machinery and equipment from the local property tax, might be turning away from that industrial land bank program, since attracting industries to the city no longer adds as much to the tax base as it once did. Is the city reevaluating the land bank program since the passage of the machinery and equipment exemption?

R: I do not think we have been reevaluating the program. I think it is fair to say that our emphasis has shifted toward a central location rather than concentrating on the outlying area. I do not think the Machinery and Equipment Exemption Act was a significant factor. That just happened to come about at the same time. I am not sure we have all the figures in yet on what the real value of industrial development is in this city or any other community. We have seen a lot of figures. I guess some of our people, some of the revenue people from the State of Wisconsin, and also officials in some of the smaller communities are seeing what we have done by attracting all this industry, and it is not really producing that much. Well, I know that there are some conflicts there in the figures, but we are still learning about fiscal balance.

QUESTIONER'S RESPONSE: We all are!

Chapter 7

PORTLAND: THE PROBLEMS AND PROMISE OF GROWTH

ERNEST BONNER
Planning Director
Portland, Oregon

In the City of Portland we have received some complaints that make us want to raise our hands and say "We have got to do something." Many times the response is "Just keep your hands out of it, and it will be all right." In many cases we have something pretty good going, but forces are pressing us. Change is coming, and growth is coming. How do we preserve what is good in the face of that growth and change?

We have an unusual government in the sense that, in theory, it works very poorly. It is called the commission form of government. We have a Mayor and four Council members, all elected at large. The Mayor assigns, by virtue of the Charter, each Council member to certain bureaus and agencies in the city. That is his basic power. One Council member is Commissioner of Parks, another is Commissioner of Police, another is Commissioner of Fire, and Water, and so on. These are doled out—sometimes for political purposes. But the result is that most of the Council has a very firm grasp of how the city operates. It has a kind of moderating influence on the clashes that always go on between political actors about how to run the city. It does not quiet down the clashes that are personality or politically motivated, but it does a great deal about getting consensus about how the city should operate.

In Portland the Mayor is very strong—not by authority, or by charter, but he is very bright. He is very energetic. He is very much interested in planning. He has supported planning to the hilt, including budgetary support and other types of support. He demands a great deal of us, and that makes a big difference. He knows enough about it to be discriminating. We do not mess around and do things behind his back.

We have a so-called Office of Planning and Development which is a modern-day contrivance designed to make city government more effective and responsive to the citizens. It gathers together the Bureau of Planning, Bureau of Buildings, and Portland Development Commission. It would like to get its hands on the Housing Authority, which is the public housing operation in Portland. I do not think it works, in this particular case. It might seem to work because of the individuals involved, but all it really does is place another layer of bureaucracy between the operating agencies and the people who decide—the Mayor and the Council. I do not think it has worked very well at all, but it is something we put up with. Frankly, it is not that important. Those who established it do not believe in it enough to actually make it work.

I think I should talk a little bit about the organization of the Bureau of Planning—not because it is crucial but because I think it says something in general about how things are for us. When I arrived, there was a staff of about thirty, organized in the following fashion: a director, two assistant directors, eight or nine senior planners, and then all the troops at the lower levels. Over the two and one-half years that I have been in Portland, one of the major accomplishments has been a new organization: a director and four chief planners in the areas of comprehensive planning, program and policy analysis, district planning (or neighborhood planning), and code administration (which covers zoning, variances, conditional uses, and related issues). The number of people is now 55 total—50 at bottom, 5 at top. We have a budget of about $1,200,000, which is about $4,600 a day or $23,000 per week. Put yourself to a real test some time. Since $23,000 comes in every week, imagine not meeting the payroll and all those people waiting with their kids at home. What is coming in and what is going out? Think about it like that sometime. I do. It is not very encouraging.

What have we been doing? Since I have been here, we have been working on four neighborhood plans: one is just about to be adopted totally—plans, zoning and capital improvements—and three others are in various stages along the way. The second has been through the Planning Commission and is going before the Council this month. The third has not been before the Planning Commission yet, but it will be in two to three months. The fourth has been through the Planning Commission and is about ready to go before the Council. Three years ago all those plans were sitting around the table; today one is almost completed and the other three will be completed before the end of this fiscal year.

One of the things that has been a problem is that there has been a lot of rezoning. Dealing with zoning ordinances is very trying. It affects everybody over a wide area. It affects the value of their property, which is what people consider the last bastion of their rights. Usually rezoning takes a lot of time and causes friction and a lot of trouble. Many neighborhood plans involve zoning trouble; in fact, many times zoning is what neighborhoods want. Neighborhoods want us to pay attention to getting the zoning fixed up so that apartments cannot come in. Then they want us to get the traffic off the streets, and they would also, incidentally, like to have the dogs stop barking.

We have also completed the downtown plan in terms of the plan itself. We have development regulations that are just about adopted by the Council. It is a matter of process now covering height, density, use, parking, and a special set of regulations for the downtown retail core designed to make that core competitive with regional shopping centers.

We have done the usual kinds of things with codes. Believe it or not, we do not have subdivision regulations in the City of Portland. We have not had them for many years. We are now in the Council with a proposed set of subdivision regulations, and they should be approved in three months. We do not have the planning and development regulations that most cities do. This is interesting about the City of Portland because it has been this way ever since they have had planning. We have had growth pressures to subdivide land, and it is interesting that we do not have regulations to govern it.

We spent a great deal of time in my early months at the Bureau of Planning with a freeway that was slated to go through the southeast of Portland, ripping out 1% of the city's housing stock, 5,000 people, and going nowhere and doing nothing except justifying another freeway, which we were not too sure about either. We spent a great deal of time with the Council fighting that freeway and then subsequently getting that freeway money turned into transit money. Last week the Council adopted a program of some $250 million in expenditures on transit projects, and it is just a matter of time before we start turning dirt on these projects. The interesting thing about that is that the freeway itself, which went through the southeast, is now being replaced by about $150 million worth of expenditures in the southeast. An additional $150 million will be spent on transportation improvements throughout the region. Politically, the transfer of that money from that freeway to transit projects is one of the best things that ever happened to us. It gave us the money to fund a lot of smaller projects that we needed much more than that freeway. In the end we will have made $225 million, and of that, $350 million will be in transit expenditures. Some of them will be very small—trying to redesign and reconstruct a street so that it works better for buses and transit vehicles. That transportation effort has been a very successful part of what we have done in Portland.

A couple of things in zoning were fundamental yet controversial. We changed the definition of *family* in our zoning ordinance to say that the family can be up to five unrelated individuals. This permits people who are not married to one another to live in single-family residential areas. We also changed a residential care facilities ordinance. This was an ordinance setting forth rules whereby social service facilities, which must be in a residential area (in a residential unit), can legitimately be in that area. Both of these affected several neighborhoods that were dead set against both of these particular programs. The neighborhoods may eventually win out. Needless to say, we relate the capital improvements program and the general kinds of things that one would do in the planning program to these proposals and others.

What does planning do? I have spent three years in the City of Portland, and we still have to get going on our comprehensive plan. This tells something about where comprehensive planning is in the stream of things—about nowhere. My own particular feeling about the situation is that it should stay nowhere. In the City of Portland, if we are to maintain some semblance of the quality of life that we have, it is going to be by understanding the forces of change and by designing the kinds of responses that will assure us that the right things are happening. We are going to grow by probably 50% in the region over the next fifteen years. In the City of Portland itself, we are going to grow by 50,000 to 60,000 people, and I think that is a conservative estimate. We are not going to keep what we have by sitting around and hoping that 50,000 or 60,000 new people will not have cars; or by hoping that some 450,000 people in the city will not have kids to send to school; or that they will not do this or do that; we cannot hope that they will not demand space and accommodations not only in their own living rooms, but also in the places where they work. Nor can we hope they will not want to drive to work in such a way that it sets demands that are impossible to meet. We will not get there by continuing to put our heads in the sand about things—for example, why do we all have to continue to go to work at the same time and do everything else at the same time? This is causing one of our most serious problems; we constantly have to design for peak hour situations rather than tolerate some congestion.

Less is enough is really where it is at. We have to get people to say that and to take upon themselves to accept that kind of response. It is very much like comprehensive planning. Comprehensive planners are like the Hare Krishna. Have you ever encountered one in the airport? They want to sell you a book. They want you to know the light. They want you to see God and Truth, and all that. If you ever walk by one of them you get an idea how people could walk by you as a planner. What will happen to us in twenty years if we do not have a plan? People could look to the comprehensive planner for the answer—but we do not know it.

I believe in it, which is not to say I think it is the greatest thing since long underwear. I believe in it, and my role is partly to do comprehensive planning for the City of Portland, and so I will do it. Yet I do wonder about it all.

Let me talk about some of the basic things that we enter into when we start thinking about the comprehensive plan, what we are doing for it, and the things that occur to me about it. The first thing is that I think we have to turn it around to a situation where we say "Let us help people help themselves." This means we do not set Utopian goals and we do not decide what is right. We do not do anything like that. We simply give people the information and the tools they need to do something on their own. If they do not do it, that is the way it goes. If they do it, that is fine, and so there is no further judgment that we can make about it. Perhaps that is the extreme of it.

As a public society, we are accepting more and more responsibility when we should be taking on less. We are not capable of discharging that responsibility, and we never were. We kid ourselves to think we can. Individuals kid themselves to think that the public will take responsibility for something they should do. Families who do not take care of their elderly mother and father are not going to survive. The family unit is disintegrating, and all the things that the family did are now being accepted by—guess who. We cannot, we will not, and we say we do not want to, yet we accept these problems as public responsibilities. When we decide that we are going to do something for somebody, we set up a whole group of things that are a lot of mischief—for example, what they should have, how we should go about it, what are our public interest goals and objectives, and other things like that. We are doing all of that in the comprehensive plan.

I have not said anything about goals and objectives because I do not think I believe in them any more. They are going to get us in the end. There is a long drawn-out process where everybody sits around, raps, and has big town meetings and even TV meetings; this includes politicians, the League of Women Voters, neighborhood groups, and so on. Guess what they concoct—nothing. That is what we get from a long set of discussions about goals and objectives. Nothing. Why would we not get that? Everybody has to be satisfied in that process. What we end up with is something that does not discriminate at all. We do not get any direction between right and wrong or good and bad. We get gobbledegook. We just cannot get anything more than that. We cannot ask a deliberative body to come to any kind of a hard decision in that abstraction. We can ask them to make hard decisions in specific situations, but we never will get them into a corner where they will say "we are going to do this or that or the other thing." They will not do it—consistently.

I would like to return to this idea that less is enough. Small is OK. The big public spending programs of the past were a delusion for us all. I think they did not do anything except convince me that we were into things that were too complex for

for any of us to understand. We did not understand them then; we do not understand them yet; and I do not think we will ever understand the complex system we were trying to deal with in big bold terms. Although anyone can see that the world did not fall apart, and we are still here, I do not think it was a good thing to be doing—nor were we successful.

Much of what we are doing in the City of Portland is in the nature of taking the risk out of investment and emphasizing the role of private investment in this whole game. The emphasis has got to be on private investment, not public, and that emphasis is taking the risk out of investment. If we stop to think about it, that is what has been happening all along. What is a subsidy but taking the risk out of investment? Then one might say, but for whom? The big emphasis in the comprehensive plan in the City of Portland will be on the neighborhoods and taking the risk out of investment in those neighborhoods by private individuals. This must be because private investment of time and money is going to be what saves them. If we do not have that, we can go home and forget it. If we do not get that, we have nothing, because there is not enough publicly gathered money in this world to do the kinds of things that a group of private individuals can do. So there is going to be a big emphasis on that.

Another thing is to start where people are, not where we want them to be in planning. Start where they are at a simple level. In this period of competition for people's interest we have to communicate with them. We have a sheet that talks about streets. It starts talking about streets by talking about front yards. Between your house and the street is your front yard, and that is how it starts. I really think that is where we have to be.

There will be an emphasis on neighborhoods and the whole relationship to the comprehensive plan for the City of Portland. It starts with a map that shows *your* house, on *your* lot, on *your* street. That is where we start. It does not start with a rap about a bar chart that shows the varied uses of energy. It does not start with a map of a regional transportation system.

It does not start with a discussion about how we control the use of land. It starts with your house, your lot, and your street. When we get there, then we can begin to get people interested. That is where they are interested, and sometimes we can get them to take as wide a vantage point as their whole neighborhood. Sometimes, among those people who are interested in their whole neighborhood, there will be a few who might even be interested in how their neighborhood fits in with all the other neighborhoods in the city. But we will not get them there without taking them from the beginning. If anything has been brought home to me, again and again, it is start where the people are, not where we want them to be. When we talk about getting them in the buses, do not talk about getting them in the buses. They have to get into the buses in their own way.

There is nothing that is so complex that it cannot be stated very simply. That seems fairly simple, but it is really hard to press. As a matter of fact, it is hard for me to get that acceptance out of the people who work for me. After we had our northwest district plan—which is the district in Portland that contains the most articulate and most responsible neighborhood organizations—and the plan was in the Council, we took the Council on a field trip. A guy from the television station came up and asked what the northwest district plan was all about. I started to tell him, and I could have gone on and on. He said "Have you got a one-liner?" Think about things in that way, because 25 years in the lives of a lot of people could be really one-liners. Some things that people spend their whole life with, you want in terms of one line. That is about as much as you are interested in, and maybe that is about as much as is important to you. At any rate, the one line is what gets on the television—at best a twenty-second spot. Put everything you have to say that is really important into twenty seconds. Get down to the important part, and get the rest off to the side.

Finally, I wonder whether or not planning could be a little more democratic. The first thing is, it is not planning, it is decision-making. It is not producing a plan, it is getting decisions made in such a way that something is assured for our children and grandchildren. Could those decisions be made more democratically? I do not know. I do not even know if they should be. But I think what we are going to try in Portland is to push to the extremes and to try to make planning more and more democratic. I am not convinced that it will result in any kind of better decisions. But it is kind of a test for planning in Portland.

QUESTION: Where are you putting the cars? You have taken out freeways, and you have taken out streets.

RESPONSE: I will be you that in the City of Milwaukee and the City of Portland it is the same thing. There are some streets that have less traffic on them right now than 25 years ago, particularly because the interstate system is there now. It is really interesting that to justify the interstate we show the reduction of traffic on the other streets, but we never do anything physically to the other streets. We could. The interstate highway had reduced the function of one street to that of serving a few special interests, mainly truckers going between two industrial districts. They molded a campaign to prevent the closing. Ultimately the head of our transportation commission and the chairman of the board of the power company said no. We have studies that show it hardly makes any difference.

Since that time, we have taken one ramp off the bridge; we are going to take two more ramps off bridges leading to the front avenue; we are squeezing down seven to four lanes. All of the original plans were just ridiculous. All of these plans had been justified in terms of traffic counts, but by and large it is a twenty minute a day problem.

The perception we have is that we want that land along the river for parks, and it does not leave much room for cars. It was the end of a 45-year effort by citizens groups, not the city, not the engineer, not the Council, not the Mayor, not the Planning Bureau, but by citizen groups that recreated themselves every ten years. They kept that issue alive until we got that combination; the need for a park and the Mayor coming together. A lot of that happens. Citizens keep after it. They were not your down-home citizens, they were very influential, but they kept at it.

Q: I get the impression that Portland has nothing in common with other cities like Detroit, Kansas City, and so on. What is wrong with Portland?

R: That may be right. I guess one of the things that is absolutely crucial in understanding this is that the practice of planning that we are going through is put in terms of our bureau and how we go about our day-to-day business. That is not very much different in any city. We have zoning, and we have a section that works on transportation planning, and we have a section that does neighborhood planning, and so forth. Many cities have the same kind of institutions, but nothing else is the same.

The attitude and approach that the Bureau of Planning has in the City of Portland is different from others. It is different because the Mayor of Portland is different. The Council is different. It is a commission form. The city itself is different. These are important differences. The distribution or income of the population of the City of Portland is not very different from the region. The distribution by occupation is in *favor* of the city. The city has a greater percentage of higher income, white-collar occupations than the region as a whole. The city has open land and we are still annexing. Another thing is that Portland is young, yet we already have 1,100,000 in the urban region and we are getting a little bit bigger. The cost of inflation is helping us, because it is keeping the relative costs of city versus suburbs more in our favor than before. The interstate is helping us. The cost of living is going to help us. We are just young, and we have not had the chance to make a lot of mistakes yet.

Q: Does Portland work on more than a 9 a.m. to 5 p.m. basis in the downtown area or is there a daily population?

R: No, not significantly so. We have some residential areas that are still strong downtown. Development regulations provide for some exclusive residential zoning so we have to build housing in those areas. I think over the next two decades that will make a difference, and we will build up some housing, but it is tough now. We are getting a few new housing interests but not too many.

Basically we do not stay alive much after 6 p.m., except on Friday nights when the kids from every high school in the state come to town. The interesting thing is how we get something started downtown. We have a policy on downtown streets. It says that on a certain street pedestrians have equal if not higher rank than other things. It means that there cannot be an access to a parking garage or a lot on that street. We cannot cut across it. It means that the community development money, as well as the tax income and urban renewal money, is going to be spent to make it a different physical thing. The sidewalks are expanding out, and the streets are going to be squeezed. Cars can go on it, but it is going to be definitely a pedestrian street, and that will be accomplished.

What is interesting about this is that once the city made that decision three years ago, that Council action meant that a tremendous number of private decisions would have to be based on that policy. The city cannot turn around that policy decision. Major new retailers have located on that street for that purpose. Now they are hollering, "When are you going to improve the street?" We have to rush to get the improvement plans.

The downtown plan started as a kind of a citizens effort among the people who were in the downtown. These were basically the retailers who had the most to lose. They started the plan in 1969 with their own private money. They got a downtown plan together, and they presented it to the Council. The Council finally set some guidelines. Since that time, we have been proposing boundaries, development regulations, specific development plans, and programs. All along it has been a private constituency wanting it with the capacity to swing it politically and enough effort to get things done.

My impression of the downtown is that it is quite large. It is difficult to make some sense out of it, but the downtown plan for Portland has some very simple and clear policies—for example, a park around the waterfront, intersecting pedestrian ways, housing, and pedestrian streets. These are very simple straightforward concepts. Now we have results.

Q: What are Portland's biggest problems?

R: We have a growing problem with police and fire unions. We do not have enough garbage collectors or street cleaners. They are becoming more and more political and nonprofessional and more and more grabby. So that is a serious problem. I think we have a serious problem in our basic fiscal capacity. It can be seen when costs go up by 8% to 10%, yet the fiscal base remains relatively fixed. I think we have a growing social problem, not so much in terms of race, but in the high schools and the junior highs. There is a racial problem. It is not severe because we have a very small minority population.

Q: Did you really mean to say that you do not believe in setting goals and objectives?

R: Yes! We could spend hours going over every one. Do we want to maintain it or do we want to improve it? Improve or enhance? Maintain or enhance? Maintain and improve? We come up with these words, and everybody who suspects there is something important about it will be there long enough to knock out anything that might threaten them. The result of all that, after many months—if not years—is nothing because there is not a single thing that can be said about anything that will satisfy everybody. It happened to our downtown plan. Goals and guidelines were established, and, with the exception of those major concepts that everyone likes, all the rest is just garbage.

People come in and want to do the most awful things: Burger Kings in the downtown, filling stations, car washes. They come in and they say it meets the downtown goals in these ways. There is this and this. Yes, but what about this and this? There is something in it for everybody. There has to be. So why do it? What I think we should do is bring them into a process that makes them in effect a constructive force in decisions.

Q: What about district plans? Do they have to be approved by each district?

R: The district plans largely emanate out of the district associations. There is a lot, maybe too much, input from them. I think that there is more than enough participation, in all phases of government. This includes capital budgeting, planning, and the regular operating budget. The Mayor is very much into giving people what they need on their own behalf.

QUESTIONER'S RESPONSE: That is what it appears to be!

Chapter 8

ST. PAUL: PROFESSIONALIZATION AND TIMING OF PLANNING

DONALD L. SPAID, AIP
Planning Coordinator
St. Paul, Minnesota

I would like to talk about how planning has evolved in the City of St. Paul, Minnesota. I am going to bore you a little with a rather long background of how it got to where it is today because I think it is important. It is very difficult to view planning in St. Paul without knowing some of the context. Minnesota, as can be learned from different papers and the general public relations documents that have come out, has a strong orientation to professionalism. As far as I can determine, this started back around the turn of the century through a grant endowment that was left to the University of Minnesota in order to educate and train people to be government servants and to deal with government in a professional manner. The final offshoot of it today is the School of Management and Environmental Affairs, and this has created a host of people in Minnesota that have, in essence, been trained to be government employees. It has created a quality of government that I find mostly unsurpassed in the United States.

We do things well. This is not just in St. Paul; this is in the Twin Cities Metropolitan Area. Now I imagine everybody who talks in this series will say we do things well, but I mean it! The state has produced a great number of major officeholders: Senator Humphrey, certainly, a Supreme Court Justice, a number of department secretaries of various offices of the United States. It seems to be quite a breeding ground for national leadership. As I get further into this conversation, I am going to make a statement contrary to that, but I am aware I am creating a dichotomy at this point. I am not a self-contradicting person, however.

This form of government has produced an innovative approach to things. Certainly the Twin Cities Metropolitan Council is an innovative approach to a metropolitan form of government; it deals with seven counties and over 100 municipal areas. It is quite a trick to bring together that number of divergent groups. The public facilities, I think, are the best in the United States, and the Twin Cities area ranks second in the quality of life. I am never sure I know what the quality of life means, but whatever it is, it is supposed to be good. We have an excellent park system. The school system is really quite good; people tend to degrade it, but it is a very good school system. The road system is quite good. A person can go from one side of the metropolitan area to the other in rush hour in 40-45 minutes. And that is with a population of two million people. Airports, the malls, the shopping centers, the skyway system or the second level pedestrian system— through which eleven blocks of downtown St. Paul are all connected together—these allow a person to live, eat, and work in downtown St. Paul, and never go outside. Such a person might die of claustrophobia, but he or she would not have to go outside. We have a number of very high-quality social programs. For example, health care; every time one of my children gets sick, I get at least two school nurses calling me regarding my daughter—is she being cared for?—almost to the point where it is a problem. But our health care system is, without exception, one of the finest in the United States. *All of this is a product, I believe, of the professional orientation of the community.*

Apathy is one word that I do not think fits the Twin Cities

area. The best show in town is the Council meeting. There are 350-400 people attending the Council meetings, and that is not unusual. We tend to judge a Council meeting by whether it was a three-bus loader or a four-bus loader. People get involved in government even to the point of selecting its leadership starting at the caucus or neighborhood block-level caucus. It is possible for anybody to enter into a caucus and end up being a delegate to a national convention. It is totally open. Labor is a very strong force in Minnesota and in the Twin Cities area. Depending on whom you are talking to, it is "Who is running whom?" It is a very strong influence.

We have more neighborhood associations than you can shake a stick at. They are very strong and very viable. We are moving toward a system of district councils. The city has been divided into seventeen district council areas, and these are slowly being recognized formally by the governing body, the city Council, as District Councils. In essence, they are becoming a form of neighborhood government; they are making major decisions that relate to their neighborhood, particularly in the area of planning. They may become the source of requests for capital improvement funds and community development funds. They review all kinds of things, such as street vacations and the like. As a result of this, we are now moving toward a neighborhood-oriented police force. The city is in the process of hiring about 35 more policemen who will be assigned on a neighborhood-by-neighborhood basis, on a walking beat, to serve as police officers for each neighborhood.

Much of what I have said has been developed over a length of time, and it is the conventional wisdom. We are now entering an era of a new awareness, but even with this strong professional background, and certainly with quality implementation with a great amount of citizen input, things seem to "go to hell" periodically. For example, there is one neighborhood on top of the bluffs overlooking the downtown that is called the Historic Hill area. It is a historic district, and it has become a very attractive area for young, married, white professionals to move into. The value of property in that area since 1974 has inflated 75%. I know one person who bought a house in 1974 for $30,000. He just moved out of town and sold it for $110,000. Now that is not bad economics! You cannot get that in the suburbs, I can tell you that! As a result, the area is limited only to those persons who have the economic capability to buy a house, and let us face it, it represents one of the better housing buys, as opposed to the new housing out in the suburbs. What a person gets per square foot is substantially a better buy. But it is displacing the traditional housing for the black and the poor. It is displacing them by using normal market standards. The normal market process does not include any relocation benefits, so it is just as forceful and just as catastrophic, frankly, as the old urban renewal approach where we hooked the bulldozer up and said "Here we come guys and girls." It is much more difficult to address because we are now dealing with the workings of the private market system. The awareness is not there. Consider an individual sitting there saying, "Gee I'd like to live here; I've lived here all my life and all of a sudden I'm being forced out by economic pressures!" There is much unrest at this point.

Where does planning fit into all this? As far as I can determine, back in the early 1960s, St. Paul had started to develop some pretty good planning. It was of an early 1960 vintage, sometimes referred to as a "701 approach." The physical plan dealing with the various functional components—the thoroughfare plan and the land use plan—was not a bad plan for that time. Between about the mid-1960s and early 1970s, planning really went into the doldrums. It pulled inward. It was attached to an autonomous board, maintained an independence from the political structure, and did a lot of nice navel-contemplating: a think-tank type of approach. Most of this was pretty useless. During that same period, planning became very much project-oriented. We had a strong Housing and Redevelopment Agency. The planning that was done by that agency, however, was essentially only the planning that was necessary to support a particular project. We would start with Block A, that is, a hole in the ground (and it has been a hole in the ground for a long time), then we would work our way out from that block. We did enough planning around that block so that we felt somewhat secure with it. Well, that was fine until we got to the next block, and we planned out from it, and then there was a conflict where the two came together. That kind of planning has been the conventional type of planning in St. Paul; it is a project-oriented type of planning. *synoptic planning?*

Why does this type of planning take place? Alan Altshuler, in his book, *The City Planning Process,* uses the Twin Cities area as an example. His case study was in the early 1960s, which is when some pretty good planning was being done. I get a kick out of it, because one of his chapters deals with the St. Paul Land Use Plan, which has yet to be adopted. He observes that getting political support for comprehensive planning in the Twin Cities area is about impossible. Part of his observation is that he perceives that the elected official sees himself as a servant of the public, and that the public also sees him as a servant of the public and not a *leader* of the public. This deals with some terms that I think are quite critical. I think with few exceptions that I agree with Altshuler's assessment that the traditional approach is that the elected official is a servant not a leader. Here are some examples of why I think this is true. The comprehensive plan that was developed in 1964 is yet to be adopted. It took from 1964 to 1975 to get the "new" zoning ordinance adopted. When that new zoning ordinance was adopted, it was ten years old, but it replaced one that was fifty years old. That had to be an improvement. It was the first major amendment since 1924. There has been no major update of any of the planning works since 1964. Now I find it hard to believe that this is a product of leadership. By its very nature, planning is directive. It feeds the fuel of leadership. It looks at things that may not necessarily be politically viable at a particular time. It has the ability to go out and look at issues that may not be issues as perceived by the voter. That is the commodity that the elected official deals in the most. He understands it the most. The last thing he needs is a bunch of planners running around drumming up business. So we do not have to take long-range views of things, and, if we can, we can avoid pointing out what may be issues coming down the road.

One of the things I always enjoy is going into a public hearing. I get killed every time. There I am standing up there saying; "Now Mr. Councilman, Mr. Mayor, now this is something that might happen around the year 2000." On the other hand, I have 300 residents saying "you are not going to put that road through our neighborhood." If I were that Council member I would say, "Oh, that silly planner!" I would vote against him, too, because that is an immediate issue. It is an immediate issue as opposed to an issue that is coming down the road. Planners have not always been right in forecasting what may happen in the future. Anyway, if he does not, he will not be in office anyway. The commodity I am dealing with is a different commodity than the planners deal with. I am dealing with votes, and planners are dealing with something called professionalism.

How do we get something to happen out of all of this? This is not only true of St. Paul; this is true of any large city or any small city in the United States. For planning to be implemented, for it to move out of academia and onto the streets where the people can get out and feel and touch and live with it, we must have a better sense of well-being, which I think is what we are all about. First, that requires political leadership, and, certainly, some political courage—that is a rare commodity. Planning must be put in terms that are understandable to the political process. What do I mean by understandable to the political process? We are working on a thoroughfare plan. This thoroughfare plan is traditional in the sense that we are talking about functions and the like. It tries to look at something called congestion, and there is so little congestion in St. Paul that it is hard to find. We have about a "Level of Service C," and we are developing a plan to go to a "Level of Service E." We are planning to cause congestion is what we are doing. That has to be a different role for me to deal with. But as in all plans, it changes the status quo.

I was out until 1:00 a.m. on a recent night on an Interstate Highway Proposal Public Hearing. There was much controversy and many emotional outbursts from various groups against the professionals. I understand that because I am a professional planner. I understand all those funny marks on the map. But it is not important that I understand it. It is more important that the Mayor understands it. The Mayor says: I am going to deal from here to here on this thoroughfare; I can understand that. I can deal with that. I understand what the issues are on that. When I get down to the end of this thoroughfare, do I have any alternatives? Am I blocking myself into a corner? My answer to him is: yes, we do have alternatives. He says: fine, then I am going to deal from here to here. You worry about the rest of the plan. That is your job. I am going to worry about this piece right here. We got it into terms the Mayor could understand. The Mayor is going out championing it.

We got into a meeting one night with about 300 irate citizens from outside St. Paul. The Mayor started out his presentation by saying that this was his plan. He said: my Planning Coordinator helped develop it, but it is my plan and I agree with it. That is the first time I heard the Mayor say it was his plan. About ten minutes into the stone throwing session, he turned around and said, "on second thought, maybe it is your plan." There is no way I can give that man a tome with a bunch of funny marks in it and a lot of traffic buzz words, which nobody really understands (including the engineers), and say: I want you to understand this; I want you to become a planner. We are going to give you a cram course so that you can go out and sell this plan. It is not going to happen that way. Let's face it, part of the process of being an elected official is conflict resolution. If there are more conflicts than can be resolved, then there is a totally impossible situation, and the plan will never get anywhere.

We had a Charter change in 1972, so I am moving away from the perspective of Altshuler's book when we had a commission form of government. We are now into a strong Mayor, strong Council form of government. We have a Mayor and seven Council members, all elected at large. Planning is located in the Office of the Mayor. The servant-leadership role idea is changing, and more and more the Mayor is perceiving himself as a leader. This Mayor very definitely does—Mayor Latiner—as did the previous Mayor. He perceives himself as a leader, but he had troubles breaking out of the past shell. We have been into it for four years now under the new Charter, and I have been through two major governmental reorganizations, and it takes eight to ten years to work. Milwaukee went through this a number of years ago, and I imagine it is still going through some of the processes of trying to sort out who does what to whom and when. I am very confident now that with the strong professional background and orientation of the community, with the new Charter and orientation toward leadership, and with the goal-setting and policy direction type of approach, we are entering into an era when planning can do some good things and be useful. To be completely frank, I have very little interest in plans sitting on a shelf. If they cannot be implemented, then I would just as soon go and find something else to do.

A number of things have happened that are helping this leadership change. When the federal government was going through the revenue-sharing issue, the *Wall Street Journal* had a beautiful article and description on why there was trouble getting passage of the revenue-sharing bill. It said that it had taken a long time for the individual constituencies and individual bureaucracies of the categorical grant programs to dig their separate and individual tunnels to the vault. They did not want anybody moving the vault. But the vault got moved! For planners, the urban renewal vault is now called

CD—Community Development. The vault got moved from Housing and Urban Development, either from a region or from Washington, to a mayor and council. Traditionally, planning has been almost a quasi-governmental agency, not having close ties to the elected officials, being responsible instead to nonpartisan commissions. All of a sudden planners had to go to a mayor and council to get money. I would hate to take a poll of the number of elected officials in this country who felt planning was worthwhile funding, at least in the early 1960s.

Dean Catanese showed me a book by Jimmy Carter in which he claimed to be a planner. In his book, *Why Not the Best,* Carter said, "I am a planner." I know a lot of politicians who would go right down the tube with that comment. Politicians should never associate themselves with planners! It is a four letter word! I maintain the reason is that the planner basically keeps the politician at arm's length. I have never met a planner who actually built something. *It is the political process that causes things to be built.*

In the City of St. Paul, we are again going through another reorganization for planning and development activities. We are in the process of trying to form a Department of Planning and Economic Development. The reason for this reorganization is essentially coming from two directions. First, a very obvious need for coordination exists. Operating agencies are going one way in the classic sense of building the street, then the Sewer Department comes in another way and cuts it up to put in the sewer. We have not done that badly, but we do things reasonably close to that. Second, a desire exists to better control policy. There is a strong desire on the part of the Council and Mayor to direct and have a grand strategy for the direction and the development of the city. There is a lot of thrashing around though with the term policy, and there is a lot more thrashing around about who is going to control the policy. Does the Mayor control it or does the Council control it? That is the dichotomy of the strong Mayor, strong Council system. Unfortunately, they are translating this into organizational terms and not into functional terms, as if an organization caused anything to happen. It is the people and the understanding of the people within that organization, not how we arrange the box on the piece of paper, that counts. Peter Drucker has an interesting idea called *Cellular Management*. He breaks up the strong pyramidal organizational structure. He is saying that in a government, the communication has to be horizontal, not just vertical. Yet our structures are designed for vertical communication, almost at the penalty of insuring that we do not get any horizontal communication.

We are approaching communications in very traditional terms. We are approaching it in department-of-development-type terms, with a planning, housing redevelopment agency, and port authority all being merged. If there ever will be a strange group, that is going to have to be the one. We are going to have sort of "environment be damned" port authority in one room; in the next room will be the "anybody got a block to be cleared" urban renewal agency; and in the third room we will have a strange group called planners who have a lot of magic markers in their hands drawing funny pictures on maps. I sure would pity the poor man or woman who will be director of that agency. It is the classic Russian Troika. But anyway it is something we have got to get into, and we are dealing with it.

Behind all that, though, is this drive to be directive, or a drive for leadership, to say we are in control of our own destiny. All we need to know is what levers to pull at the right time. I submit that this has to be one of the best climates in the world for planning to enter into the picture. There is a vacuum here, and planning should fill that vacuum.

I am dealing with the last two years, which is a long period of time when there was very little planning moving fast enough to fill in some of that vacuum. We are very much in a position of playing catch-up football. I am the world's greatest opportunist at this moment. *I am convinced that planning implementation is a product of timing.* If we do not have that plan and that planning component done the day before the decision needs to be made, then we are at least one day too late. So we are now knowingly dealing with incomplete planning. We do a piece of this and a piece of that. We do a multi-service center plan; we really have not looked at the entire delivery system for human services, but we make a best bet that human resource delivery can be most effectively delivered through a multi-service center approach. We may be proven wrong as we get into the analysis of delivery systems, and we may be criticized for what we should have done first. But that is called opportunistic timing. We will have to change it some day if we are wrong. But again, there was a great need for some kind of direction on how to deliver human resource services. We entered the arena and took our best shot. If we did not do it, somebody else was going to do it. We figured we were at least as qualified to make that guess as was somebody else. So we did it.

This may not quite fit the classical planning process mold. We cannot bring all elements of a plan along equally. For example, we are working very strongly in the area of housing. We are spending about $9 million a year in St. Paul on housing rehabilitation. Obviously, we have to get a little further ahead on a housing strategy, and on a thoroughfare plan, or maybe on an economic strategy, simply because that is where the resources are being spent today. Again, we are being opportunistic at this time. We are using our best judgment because we have not completed all of the nice data-filled studies.

There is a book by Fred Bair and Richard Hedman called *On the 8th Day.* One of my favorite cartoons in that book is of two planners standing with computer printouts up to their necks. The caption reads, "Do you think we've got enough data to make a decision?" Frankly, I have only been surprised by data once in my life. Usually when data tells me something I do not know, I question the data. I know that we are doing a lot of hipshooting, but again we are doing it because we are in the political arena. If we were not in the political arena, we would be in the back rooms doing our nice "matrices of environmental parameters."

The end product of planning, then, must be a development strategy. To me, if planning does not end up in strategy terms, it has fallen short. It is only through strategy terms that the translation to the elected official can be made. If we stop short of that, we are not going to be effective. At best, we might be able to get an article in the *AIP Journal.* I stated earlier that the city had developed a very strong ability to implement and did enough planning to support particular projects. That is the conventional wisdom. I am not going to try to change that conventional wisdom. It has taken a long number of years for it to arrive at this point. So what I have to do is translate comprehensive planning into project terms. I think it is easier for me to do that than to try to get the whole thought process of the community translated into other kinds of terms.

The way we are approaching this is by becoming strongly involved in development programming. We are not going to get it down to a sequence of steps to implementation, such as when to have the cement truck arrive on the site. I do not mean that level of detail. We are going to get into a level of detail on what parcels of land are to be bought and in what sequences, when does the thoroughfare have to be built, where is the right-of-way for that road, and so on. That is when we start construction of housing—only when the sewers are already there.

Dealing with it from a comprehensive planning and strategy viewpoint is difficult. I really do not know anywhere in the United States where this kind of approach has been carried out. I know where various pieces of it have been tried, but I do not know where it has been done on a comprehensive basis. The City of Richmond, Virginia, is trying this approach by using a city development team. Each year they meet under the direction of the Planning Department, and all the operating departments try to bring together the activities that they are going to carry out that year in a coordinated manner so they become mutually supportive.

We are going to try to do it by laying out a kind of blueprint. We are going to try to take the comprehensive plan and run it through a development strategy, which means we are going to be getting into priorities. The Chairman of the Council Finance Committee said that by next year we want the capital improvement program to tell us things like housing is 20% more important than thoroughfares. That is frightening to me from the standpoint that it is the kind of thing planners have always said they wanted but also wished they never had. We are going to be put into a put-up or shut-up situation next year. I find that exciting. I love mountains to climb, and that must be Mount Everest.

I think it is exactly where planning should be in a community. It is in a leadership role. It is being translated by elected officials but not replacing elected officials. Can you imagine living in a city that was run by planners? It would be just one big AIP conference. The development strategy needs to be nurtured, looked at, and qualified by the elected official. Planners always get nervous with elected officials for some reason. They seem to think that if we allow an elected official to pass judgment on something, that is going to compromise our plan. The plan is the most important thing! The inviolable plan. One of my favorite analogies to that is that if we are going to have a baby, I do not know of any other way but to climb into bed. If we are going to have a city that is developed in accordance with the plan, the planner and politician are going to have to get in bed together. It is just as simple, and as gross, as that. In my experience I have found that a very rewarding experience.

I had only one experience in my entire life when an elected official said, "We want you to change your mind." He did not give me any reason why. My answer to that was simple—no! Later, he went away. There have been many times, however, when an elected official has changed my mind simply because I was wrong. An elected official has an entirely different perception of a plan, or what the planning process should be, than I often do. He is seeing it from a different viewpoint. The only thing we have to do from a communications standpoint is to get the same perceptions.

I think the same thing is quite true of this term "citizen participation." I love citizen participation. I frankly feel that is where it is at. An elected official deals with citizen participation every day and certainly at election time. For a planner to sit back and keep the people at arm's length is silly. It means that we are going to have to change our approach. We certainly are not going to use such terms as "matrix of environmental parameters." We had a discussion once about the reading level at which planners should be writing, and I am sure that if we are not writing to a sixth-grade reading level, we are not talking to anybody other than our professors (and we may not even be talking to them). I find that a lot of people write beyond their own grade level, so all we are trying to do is create a big snow job. When we get out into the community, somebody in the back of the room will say "What do you mean?" and that is exactly the problem. We are not talking, and we are not communicating.

I have found the genesis of a fulcrum for the thoroughfare plans in St. Paul. I found it simply because I was drinking beer with a couple people in a neighborhood bar. One little lady was talking about the day when they would build that road and when she was going to lay down in front of the bulldozer. We got to talking about why not get in the seat of the bulldozer rather than in front of it. I started thinking about what these statements meant. I found that some people had excellent reasons why and why not in a planning context. They were talking very much in a neighborhood context. They did not want the noise. They did not want the displacement and all that. In looking further at that issue, I found that, in a planning context, it indeed was the wrong place for a road. It was not going to serve the people we thought it was going to serve. It is now out of the system, and it caused a complete reordering of the thoroughfare plan that I am going to be recommending to the City Council. It all came from that little lady in a neighborhood bar. She has lived there all her life, and I do not know what kind of education she had, but she was a heck of a lot smarter than I was.

I think that planners have to address that kind of thing. It is going to make our job much more difficult. What we are going to have to do to develop our plans is to talk with people before those plans are developed rather than afterwards. I would submit that in the time frame of things, it will probably take just about as long—in fact probably not as long if we have strong citizen input in the beginning—from the time when we start the plan to the time when it is on the ground. Again, that is how I measure things. I measure planning in implementation. If I measure it only as when I start the plan and when I publish a report, I can be finished really quickly, but it is useless.

In the seventeen districts, we are, in essence, doing comprehensive neighborhood plans for each. The actual technical work of those plans is at the most sixty working days from the planners' standpoint. Most of them take between one and two years, some of them as long as five years, to publish, simply because we have to move at the same pace as the citizens within that neighborhood. What happens with this kind of thing once it is developed, once there is strong support on the part of the citizenry, is that it will survive the political process; that is where our continuity in planning occurs. The continuity in planning lies with the citizens themselves who understand that plan, who will insist upon that plan, and who will always be there to advocate it in times when decisions are to be made. This is probably the best capability for planning implementation there ever could be. We get early warning signals when a plan needs to be turned around.

I have been talking about what I think are some new roles for planning. I am not sure how multiple regression analysis fits in. I think there is a great time for "ad hocery" on the part of planners, but there constantly has to be a firm understanding of the planning process, even though it gets bent every once in a while. I describe it as the Amazon River. Comprehensive planning is charting the main channel, but there are a bunch of tributaries. Periodically, we will be forced to go up one of those tributaries, but we will always know where the main channel is to get back down to the end point. Once in a while the whole river gets turned around, but we will know when that river has been turned around. The Potomac River has to be running upstream right now, and there has to be a good reason. This may turn around, and we will get the Potomac running back downstream again, and urban areas will have a full partner from the federal government once again. This is what I mean by a change in direction that the comprehensive planning process leads toward.

QUESTION: How does the St. Paul City Planning Commission fit into all of this?

RESPONSE: The City Planning Commission is independent of me. I do not work for the Commission. Under the city Charter, a recommendation passed on to the Mayor and the Council by the Commission has to contain my recommendation with it. My recommendation could be different. I am an independent body in that sense. I like that relationship very much. It gives diverse reviews of any planning product through the Commission that is charged with that responsibility. It becomes a checks and balances system to the planner. Yet, if that checks and balances system moves too far in one direction, then I still have the right and responsibility to give the Mayor and Council my professional opinion on it.

Since I have been in St. Paul, I have never found an occasion where I had to send in an opinion different from what the Commission did. There was one time when I thought I would, and I think this is good. If I were a mayor of a city, I would want as many opinions coming at me as I could get, and I would like to know where those opinions were coming from. I would want a Commission that was a diverse group, and it would form diverse opinions. I would want a planner who would give me his professional opinion. I would know where they were coming from, and I could judge and evaluate those recommendations. If only one opinion of planning comes to a mayor, and if that is the only way it can come, then the mayor does not know much.

Q: Do you feel that you and your staff should develop an overriding goal or objective? Do you feel, if that is to be done, it should in fact come from you and your staff, or should it come from the Mayor as a directive, or should it come from the Council, the Planning Commission—who should instigate that kind of directive?

R: We are in the process of actually having the formal adoption of an overall goal and objective for the City of St. Paul. It evolved from a study that was done by Hammer-Siler-George Associates. They were asked to look at the economic policy of the city. Economics has been a big thing lately, and one of the things identified in this study was that neighborhood stability was vital to the City of St. Paul in a real economic sense. St. Paul is only as strong as its neighborhoods. The professional planners put together a policy statement mainly to direct the allocation of capital resources. *The overriding goal of the policy statement has two perspectives: neighborhood stability and economic viability.* Then it gets broken down into substatements. This has gone to public hearings and it has gone out to all of the neighborhoods for discussion. Originally the document recommended neighborhood stability only, but now economic viability has been added as a result of these discussions. It has been adopted by the City Planning Commission as a segment of the comprehensive plan. It is now being reviewed by the City Council, and we believe it will be adopted. So it will become an official policy of the city and will become the major thrust of planning direction. That is nothing new in that it has been going on anyway. All we did was to put down in writing what it was we were doing.

Q: I read that in 1972 your Metropolitan Council instituted one of the earliest Fair Share Housing Allocation Plans. Can you tell me how successful it has been in dispersing low and moderate income housing throughout the metropolitan area?

R: The Housing Allocation Model and the distribution of housing units, particularly Section 8 units, went into effect during the winter of 1975. In essence it allocated about 80% of the units of low and moderate income housing to suburban areas and about 20% to the core cities. Philosophically, I have absolutely no disagreement whatsoever. The core cities, if we keep going the way we have been going, would become the repositories of housing for the poor and elderly with very little freedom of choice on the part of these people. There is only one big problem. Nobody at Metropolitan Council ever figured out how it was going to happen. Putting up a table with a bunch of numbers on it does not get the job done. It does prevent the cities from receiving the funds from the federal government, which they need to put into low and moderate income housing, and thus effectively reduces the overall supply to the low and moderate income people least able to cope with the situation. The net effect is less housing for the poor and elderly. The Metropolitan Council has since created a metropolitan housing and redevelopment agency, and they are now producing a metropolitan mechanism to implement that. One of the major arguments we have is that we prefer to get at it incrementally, so that while we are doing it we are not curtailing housing supply for those least able to cope with the situation. We lost. In these kinds of debates, it is an all-or-nothing situation, and nobody can say, "Let's do it in bites."

We are now starting to get low and moderate income housing into the suburbs, and it is moving along very well. They have about forty suburban towns that are applying for the units now, and it is off and rolling. Those first couple years were kind of grim. So, in effect, whether they agreed or disagreed, it was a transitional process, whether anybody wanted to say that overtly or not. We could have saved an awful lot of agony for a lot of people if there had been a little thinking on how to process that planning.

Q: To what extent are you capable of planning independently, or to what extent does the planning in Minneapolis have a major impact on plans that you would devise?

R: Obviously there can be a significant impact. There are some geographical barriers, such as the Mississippi River that gives some topographic division. I meet on a bi-weekly basis with my counterpart in Minneapolis. We kid each other about meeting on the bridge, and he gives me the visa the next week. On the housing question of Fair Share Allocation, he and I sat down and worked on a proposal to the Metropolitan Council. From that it has now spun off, and it has been accepted by the Association of Metropolitan Communities. We are now getting the market study that is necessary to support this housing allocation model. This came because Bob Moffit, who is my counterpart in Minneapolis, had the same problem I did, and we sat down and worked it out together. When we think we are going to have a conflict, we usually pick up the telephone. It was decided that, for example, on a downtown people mover system, St. Paul was the only city in the State of Minnesota that would submit an application to the federal government. Minneapolis discussed it and decided they did not have as good a shot at it as we did. So we do work together. Every once in a while, though, we have our little disagreements.

Q: There has been state legislation, as I understand it, that would require a "Mini 701 Program." The Metropolitan Council was supposed to do a level of policy planning and each community was supposed to do some functional plans.

Then it is supposed to be integrated in some marvelous fashion. How does that affect you?

R: Rather severely! There is an act that was passed, referred to as the Mandatory Planning Act, that requires all cities to develop a comprehensive plan and have it adopted by the governing body and by the Council. This plan is to include the methods for implementation and also is to be adopted by the governing body. That is going to be fun! Before we get started on this, however, the Metropolitan Council is supposed to give us what is called a Systems Statement. The Systems Statement is supposed to be bigger than a breadbox and smaller than a house, but it looks like it is bigger than a breadbox and smaller than the superdome. That is, it is pretty broad! In fact, it is so broad that it is not going to be very directive. What is going to happen by not being very directive is that each city is going to develop its own plan. It then will submit that plan to all the adjacent cities and school organizations. They then have six months to review those plans, and it then goes to Metropolitan Council. The Council has another three months to look at those plans, bring them together, and do whatever they need to do. Then there is a sixty-day period for appeal. We are up to eleven months now, and then it can go to court. It is going to make the plan adoption process awfully difficult. So I am hoping that the System Statement coming from Metropolitan Council is going to be more directive, because with the current approaches I am sure we are going to go right by each other. The idea is good. Unfortunately, there is not a commonality of perception that planning is a leadership tool among all the cities, and this is going to cause real donnybrooks down the road. I think the objective and the idea of the Mandatory Planning Act is good, but it is not a panacea.

QUESTIONER'S RESPONSE: We were hoping for a panacea!

Chapter 9

LEARNING BY COMPARISON: LESSONS FROM EXPERIENCES

ANTHONY JAMES CATANESE, AIP

Dean, School of Architecture and Urban Planning
University of Wisconsin–Milwaukee

What have we learned from these talks with planners? Is there any evidence that these talks represent seven unique experiences, or is there a basis from which we may generalize about the state of urban planning in central cities in the United States? We believe that the latter is the correct interpretation, and we have tried to select and present these talks with this in mind.

We have tried to select the chief planners from central cities that were within a similar population range, especially with regard to the urban region that they serve. We have attempted to select successful planners who are inherently professional in their experience and style. Yet we have attempted to select central cities that faced different types of problems taking place within decidedly differing environments and socioeconomic contexts. Our initial fear was that these talks would be somewhat repetitive and overlapping due to the similar backgrounds of the planners. We quickly discovered that these planners behave and practice in accord with the environment within which they are working and the constituencies that they serve. It became apparent, as well, that the personalities and styles of these planners were major factors in their approaches to planning. We believe that the reason such planners are effective, within these different cities and environments, is precisely because they have the proper personality and style for the planning situation at this time.

THEORY AND APPLICATION

One of our original concerns was the relationship or applicability of the theory of city planning, as defined by the literature of the field, to the actual practice. We asked all the planners to comment on that question, and we were somewhat surprised by the responses. We suspected that there might be a gap between theory and practice—we found instead a chasm. There appears to be little congruence between the theoretical works in the field of city planning and the practice that we have been examining. Most of the planners we talked with had little interest in the detail and intricacies of city planning theory, although they were aware of the major writers and overall themes. The specific content and breadth of that body of theory seemed to hold little relevance for these planners.

This is an intriguing finding. Does it signify the often-mentioned academic and impractical bent of theoreticians, usually based in universities or research institutes, or does it allude to the reality of the practice of planning, which may not have the leeway for experimentation? The planners in this set of discussions were clearly of the opinion that the theories were unrealistic or esoteric. Almost all of them had a deep skepticism and, in some cases, cynicism for the theories found in the literature of the field. This skepticism was often focused on the substance of the theory, but in many cases it was focused on the person propounding the theory. It was usually stated that the theories and theoreticians were divorced from the realities of planning practice in large central cities and that the persons writing the theories seemed to be out of touch with local politics and politicians.

This latter set of complaints was not new. What was somewhat different was the extent to which this group of planners was concerned with the pragmatics of city planning, especially as related to the successful implementation of planning. Many said that they only used practical and simple rules of operation to insure that implementation would be possible. This was typified by Carroll's displeasure with all but the nitty-gritty of practical planning implementation. In essence, we may have noted a trend toward pragmatism in city planning and away from the intellectual orientation of previous times. If that is in fact a trend, we may have unveiled a move toward pragmatism as the basis for planning theory. We find this to be somewhat surprising, because one would not likely reach that conclusion from the literature of the field. Perhaps talks with planners, such as these, are more reflective of this trend than is the existing literature.

An interesting aspect of the applicability of city planning theory to practice has to do with the theory of planning as a process. By that it is meant that planning is a basic process of analysis, which can be described as a series of steps or actions. In its broadest sense, the process of planning could be described as a way of thinking that is inherently regular and discernible. The results of our talks with planners seem to clearly indicate that there is a process of planning that is not too different from the prevailing theory. Most of the planners in this work were conversant with a basic procedural approach to planning that did not differ from city to city.

To be specific, it would appear that we could synthesize a process of planning from these talks with planners to include the following steps:

1. Define the basic goals and objectives
2. Ascertain problems that impede attaining same
3. Formulate and articulate alternative courses of action
4. Evaluate these alternatives for priority ranking
5. Make recommendations to decision makers

6. Assist in the implementation of the alternative selected
7. Monitor progress and provide feedback.

If one were to read extensively in the current literature of the city planning field, this basic process of planning would be observed most frequently. This process may even be definable as "classical" city planning theory, and, hence, we can conclude that this branch of theory is widely applied.

Much of the application of the process of planning, as well as the little application of theory that may also occur, appears to be undertaken by persons subordinate to the chief planner. The planners we have talked with are at the top of their organizational hierarchies and appear to rarely perform all the specific steps in the city planning process. The whole *mélange* of other planners in city planning agencies—principal planners, senior planners, junior planners, transportation planners, health planners, and so on—appears to be the group that actually does the work, usually technical. The chief planner will often set guidelines for technical work, and sometimes even refine the results for presentation purposes to decision makers, but usually will not personally do the technical analysis and findings. This characteristic appears common to all of the chief planners in this group and may imply that there are hierarchies of functions that are coterminus with organizational hierarchies in city planning. The general finding would seem to indicate that technical endeavors decrease directly with the rise in position in the city planning hierarchy.

Perhaps the clearest evidence of this characteristic is the curious manner in which most of these chief planners discuss the day-to-day functions of their agencies. It was common to hear one of the planners say; "Of course, we do things like land use and zoning studies, too." That would seem to show a feeling for a vast array of functions and activities that may well be self-sufficient. The typical functions and activities of a city planning agency would certainly include at least the following: data collection and analysis; mapping; special projects; population and economic studies; land use inventories and projections; traffic counts and analyses; and development controls such as zoning and subdivision regulations. The apparent role for the chief planner may be that of management-by-exception or management-by-objective. In other words, the chief planner does not really come into contact with these daily operations unless there is a problem or some special reason to use one of these operations to reach a goal or objective. We believe that the reason for the widely accepted theory of the city planning process is precisely to provide a basis by which these activities may be related in plan-making and continuing planning. The city planning process becomes the glue by which these functions and activities, as well as their products, can be held together. The chief planner provides oversight and guidance, but more technically oriented persons provide the services.

Our conclusion about the process aspect of planning is that it does exist in theory and practice and, as such, appears to work well and provide the underpinnings and foundation of the city planning movement. This means that city planning is several things. It is a movement when examined at the highest level of application where most of the planners we have talked with operate. That movement appears to be more managerial than reformist, and it appears to be concerned with reaching social, economic, political, environmental, and physical goals and objectives. Yet city planning is also a technical process with a prescribed set of steps with supporting techniques and methods. Both the movement and technical process nonetheless take place within a political setting subject to wide public scrutiny. The success of the planning director would seem to hinge upon the achievement of a proper balance of technical presentation and political astuteness. None of the planners in this series seemed to think that technical merits would always win out, which certainly confirms most of our experience.

This brings us to the clearly discernible characteristic common to all of these planners—politicization. All of the planners in this series made it known that they were working for political leaders and within a political setting and, hence, had to rely upon successful performance in the political arena. It was quite fascinating to hear so many chief planners from such diverse cities explain how they tried to make the nontechnical skills of the politican useful to planners. For example, there was much talk about compromises, forming coalitions, and predicting the political feasibility of plans. While this may seem eminently reasonable, there is very little such discussion in the literature of planning (and what exists is very recent in origin).

There is most likely a correlation between the high level that most of these planners occupy and the politicization characteristic. There was some commentary implying that the chief planner did not like the subordinate planners getting involved in political disputes. That may be the reality of hierarchies in local government, and it is probably not an unreasonable stance. The chief planner appears to be getting more and more involved in the political process, including partisan politics, because there is widespread recognition that decision-makers take political factors into consideration at least as much as technical merits. This means that top city planners are acting with this in mind when the major thrust of a city planning agency is toward implementation of planning rather than only plan-making.

This may signify a long-term trend or at least show that city planners are now more willing to discuss such matters. It was not uncommon for the predecessors of the planners in this series to avoid involvement, participation, or discussion of matters that could be considered political. The argument was that city planners were professionals and, as such, apolitical. The apolitical professional relied upon civil service tenure for survival and technical merit for accomplishment to the exclusion of political involvement or influence. Apparently we can judge from this series of talks that there is much skepticism toward such politically neutral practice. Even when Spaid was discussing the high level of professionalization in his city, he went on to make it clear that, "it is the political process that causes things to be built." What would seem to be countervailing, politics and professionalism, is really quite compatible in the minds of these planners. Whereas a decade ago a person was thought to "have left the profession" when accepting a political appointment, it now seems comfortable in the minds of this group of planners to be both professional and politicized. We are not so naive as to believe that other planners have not been involved in similar political activities, but what is far more interesting is the openness and frankness with which the contemporary city planner can discuss such matters.

The writer of this chapter once wrote a book, *Planners and Local Politics: Impossible Dreams,* in which two types of professional planners who engaged in politicized practice were described as *Covert Activists* and *Overt Activists.* The Covert Activist was a planner who remained inherently professional but was aware of the political process and the way that it affects decision-making for planning. That person would maintain a low profile but nonetheless utilize the political arts when necessary. The Overt Activist would be a person with a higher profile in local politics and might even be the mayor's "right hand person." We think that both styles of behavior were visible within our group of planners, and we will leave it to the reader for classification. More interesting, however, is our belief that a third type of planner, defined in the same book as the *Apolitical Neutral,* or a person who sought to maintain a technician stance with no leeway for political compromise, does not appear to be applicable to any of the planners in this work. Perhaps that is predictable since these are well-known and successful planners—hardly faceless bureaucrats buried in local governments. Yet there was no loading the deck when we selected these planners to talk. We prefer to think that the successful city planner is the one who can transcend pure technical content and blend in the political realities needed for implementation of planning.

One further point should be made concerning this characteristic of politicization—that is, the demise in significance of the "independent planning commission." The history of city planning in the United States really starts with the late nineteenth century and early twentieth century Reform Movements. These led to the peculiar notion of keeping city planning sacrosanct, protecting it from evil politicians by having planners work for a righteous group of public-spirited individuals who composed an independent planning commission (which was usually called the "city planning commission"). The reformers believed that this would allow planners to deal with substantive issues of city growth and development in such a manner that only the objective and nonpolitical aspects of the matter need be examined. In essence, the commissioners shielded the planners from the politicians. To a certain extent, the commissioners thought of themselves as the voice of the people.

All of the planners that we have talked with in this endeavor have abandoned any pretext of belief in that early theory. Some of the planners expressed an arrogance toward the independent planning commission, but most viewed it as a curious relic from the past. One wonders why we bother to have such commissions at all given this low assessment of their efficacy and the shaky foundation of their *raison d'être*. All of the planners in this discussion have been oriented toward city planning as an arm of the executive branch of local government. They see the chief executive, not the city planning commission, as the representative of the people. This is certainly the mainstream of contemporary thought in city planning, and it shows that these planners have emerged from the isolation and aloofness sometimes forced by reporting to such an independent planning commission.

We take this as a healthy sign and further evidence of the compelling urge by these planners to see more planning implemented. About the only value for the independent planning commission to emerge from these talks was a dubious feeling that, sometimes, "we need their support for critical issues." Yet there is little to indicate that such backing is sufficient to change things. Similarly, there was a suggestion that the commission still performed somewhat in a protective manner when a city planner was under attack. But again, we heard nothing to indicate that it made a difference. There may be some good reason for keeping an independent planning commission—whether it is historic or not, status symbol or not, or even if it is simply a way of getting a few more influential people involved in local government—but a more significant conclusion is that that archaic theory of the independent planning commission is inoperative.

A final source of discussion concerning the theory and application of it in city planning has to do with our attempt to postulate an overall frame of reference for each of the planners in our endeavor. We sought some device by which we could garner an overview or characterization of the city planning practiced by these planners in their respective cities. The device we employed is debatable and perhaps controversial—we even realize that it may be a bit unfair. We simply asked the city planners we talked with to describe the *overall goal* of their planning. That request is debatable because planning is a complex undertaking, and there may not be a single or overriding goal in a given city. It is controversial because it attempts to put planning theory, assuming that a goal is related to a theoretical base, in the form of the "one-liner" that Bonner described and extolled. We realize that these one-liners may "get on the television," to again use Bonner's words, yet they are just that—headlines. It may even be unfair to ask for the exposition of an overriding goal in that a planner may not have thought about planning in such a framework.

We did it anyway. We asked the planners to describe the overriding goal for planning in their respective cities. All of them did it without hesitation or protest. We were frankly surprised by this and had expected some protest. This leads us to suspect that Bonner may be onto something important when he says, "there is nothing that is so complex that it cannot be stated very simply."

The statements on the overriding goal for planning in each city by the respective planners are arrayed in Table 9-1. We are not entirely sure of the significance of such statements— or the utility. We are certain that such statements make for fascinating discussion. It also allows a level of generalization and abstraction that is most interesting in comparative studies. Consider, for example, Krumholz' reasonable statement about giving a wider range of choices for those groups of people that have the least degree of choice in their urban lives. That does not appear, *prima facie,* to be a radical goal for a city, and, consequently, city planning may be guided by such an overriding goal. Yet, like the Cleveland program itself, this is indeed an extraordinary statement by a chief planner. It clearly defines a framework and set of constraints that give a basis for specification of daily city planning activities. It defines a specific set of clients for planning, and it may even exclude some potential and traditional clients. The goal is a social one, and it would imply that this quest takes precedence over other goals, at least in the sense that it is expressed as an overriding goal.

Bonner's statement was quite different. With no aspersions to pop culture or West coast governors, he said that "less is enough." That is less of a slogan than it is a framework. Bonner seemed to be saying that planning in Portland is directed toward a new philosophical perspective that has surrounded the growth control movement. Portland has a planning program that is guided by a goal of minimizing the impact of new growth and development even to the extent of discouraging it. While consistent with a long-time concern of the State of Oregon to discourage new in-migration and achieve a zero population growth, it is the embodiment of a step-further effort to make do with less through planning. By way of contrast, it should be noted that Bonner once worked for Krumholz in Cleveland. We suspect that Bonner is expressing a goal that is entirely reflective of the middle-class values

TABLE 9-1

Overriding Goal Comparison

Eplan—Atlanta	"to make the city tolerable to live in . . . the basic common denominator there is to design a city where you can raise a child . . . and to help guide the decisions . . . toward a common goal . . ."
Krumholz—Cleveland	"promoting a wider range of choice for those Cleveland residents who have few if any choices."
Carroll—Indianapolis	"input into the goals-setting process . . . [and] budget-making process . . ."
Vitt—Kansas City	"The overriding goal of the department is to help the city as a whole and each subarea . . . work toward performing closer to the potential that exists . . ."
Drew—Milwaukee	"We decided that fiscal balance would be our overall goal and the cornerstone of our comprehensive planning and programming efforts."
Bonner—Portland	"Less is enough is really where it is at."
Spaid—St. Paul	"The overriding goal . . . has two perspectives: neighborhood stability and economic viability."

and mores of Portland, which is not at all like the goal to which he directed his activities in Cleveland. We do not believe, in other words, that such a goal could work in Cleveland because it does not provide more choices but fewer. This reduction of choices may work in a middle-class environment, but it would seem dubious in a depressed and low-income central city.

Vitt expressed a goal that is close to Krumholz' in a taxonomical sense. Vitt said that city planning in Kansas City was directed toward a goal of helping the city and its neighborhoods reach their potential. This would seem to be consonant with Krumholz' approach, and, perhaps, mildly conflicting with the approach of Bonner. There is a distinction, however, in that the Kansas City approach is invariably pro-development and supportive of growth. Less is enough is not good enough in Kansas City. Vitt defines the potential of development and growth in the subareas and tries to help concerned groups reach that potential. In the process, he believes that economic and social benefits will flow. In this way, he may be going toward the same goal as Krumholz in an entirely different way. While Krumholz directs his work to the underprivileged, Vitt directs his work toward the privileged who can build things. Vitt is as concerned with these people of little choice as is Krumholz, but he prefers to rely on classical theories of trickling-down benefits and overall lifting of life styles through growth and development.

We were pleased to see that at least three of the planners identified economic and fiscal matters as overriding goals for their planning. Drew said it most clearly when he described Milwaukee's decision to make "fiscal balance . . . our overall goal." That makes it quite clear that planning will be directed toward development that will produce a favorable cost and benefit ratio. Developments that pay their way through taxes and economic upgrading will receive a high priority in Milwaukee's planning program. This was similar to Spaid who said that St. Paul had an overriding goal of economic viability. He differed from Drew slightly in that he saw this occurring largely through neighborhood stability. Drew was more concerned with economic development due to the industrial character of Milwaukee. Carroll was concerned as well with these matters, but he was particularly interested in economic and fiscal goals that could be attained by better involvement of city planning and budget-making. This attitude was echoed by others, especially Eplan.

Eplan stated an overriding goal that was different in flavor from all the rest—indeed it may even smack of romanticism. Eplan said that the overriding goal of city planning in Atlanta was to make that city tolerable to the extent that it was a fit place to raise a child. That is a fascinating statement! It points toward a notion that is often overlooked in city planning—that of life style and quality of life for both the middle-class majority and all minorities. Eplan describes a goal that would seem acceptable to all groups and classes. This goal has wide-ranging implications for the economic, social, and physical form of the city. Yet it is an understated device for guiding planning. It is an attempt to establish a design technique that one may hope will produce economic, political, and social benefits.

The goal described by Eplan is broad. It may be difficult to implement in a given situation. Eplan deals with this problem by focusing his planning activities largely on neighborhoods and budget-making. Thus, he is able to interrelate the romantic with the pragmatic. By giving such a concentration to city planning, he expresses an approach that may be successful. This approach seems to avoid some theoretical problems that might exist in other cities. Eplan can avoid criticism of class favoritism, since that is only remotely implied in this goal. He does not have to take sides in the growth control versus development or environmental versus economic issues unless they have implications for the tolerability of the city. Eplan seems to have stated an overriding goal that manages to be classless and impartial yet filled with hope and confidence for central cities.

INNOVATIONS

Another area of inquiry that we raised during these talks with planners was innovation in city planning practice. Flowing from our initial concern with the application of planning theory, we sought to identify practices or programs that were not strictly theoretical applications but rather that were manifestations of a particular city or set of circumstances. We also were interested in innovative practices that might be instructive for transfer to other cities.

There appear to be a number of such innovations, in our opinion. Perhaps the most striking is the sincere and major commitment and accompanying set of practices and programs for the preservation, restoration, and improvement of neighborhoods in the city. Such innovation was most clearly seen in Atlanta, St. Paul, Milwaukee, and Indianapolis, but also strong signs were apparent in Cleveland, Kansas City, and Portland. The reason we use the word striking to describe this innovation is because it is not widely heralded in the literature of planning (although some of the popular press has featured such programs).

There has long been a concern with neighborhood planning in city planning practice. In more recent times, much of this concern has focused upon the *neighborhood theory* of suburban development as well as the standards-satisfying approaches linked to the public health movement in America. While a commendable movement in the post-World War II period, there were troubling consequences, not the least of which were residential and school segregation and urban sprawl, all of which contributed to central-city decay. In 1961, Jane Jacobs wrote an indictment of city planning practice, which she entitled *The Death and Life of Great American Cities*. She criticized the trend of city planning to abandon the older, existing neighborhoods in favor of newer developments with their acceptable if antiseptic standards. She challenged planners to rediscover neighborhoods in order to bring new life to these urban places. There were several later works both supporting and attacking her challenge, yet she won popular approval.

It appears now, over fifteen years later, that Jacobs' challenge is being met in several cities. There was much talk about the return to urban neighborhoods and the impressive innovations in planning practice to make such areas "fit to live in," to use Eplan's phrase. Eplan wondered whether this was a real return or a new discovery by the younger people of the richness of urban living. When one considers that this younger generation may have been raised largely in the suburbs, never experiencing city life, he may be seen to have a valid point. Other contributing factors that appear strong in virtually all of our cities under discussion would include: high gas and commuting costs, rising new construction costs, the sterility of cultural and social conditions in suburbs, and the spread to the suburbs of crime rates previously considered to be a central-city phenomenon.

That there is a regeneration in all of these central cities in some of their older neighborhoods is without question. This is not a trend that can be described in statistics, but rather it is one that has qualitative measures. There is no rush to repopulate central cities, in other words, but there are clear signs of urban neighborhood living by a higher income, professional class of younger people.

The city planning programs discussed by the planners for Atlanta, Indianapolis, Portland, and St. Paul seem to place the highest priorities on planning for such neighborhoods. However, in all of the cities surveyed there was a significant set of innovative programs. The typical approach appears to be one where the city planning agency helps to organize the neighborhood and provide technical services. Many of the planners stressed that the success of rejuvenation rests largely with the strength of such neighborhood organizations and their leadership. Bonner made it clear that planners should not attempt to create leadership where none exists and may very well have to concentrate on the strongest neighborhoods. Eplan discussed the real differences in neighborhood leadership and why he was satisfied with survival and success by the fittest. Spaid seemed to be saying that the essence of city development was neighborhood development, but it was a matter of political leverage for neighborhood improvement.

While these may seem like tough statements from public planners, we were more impressed with the *Realpolitik* stances. These city planners seemed to be saying that the advocacy approaches of the 1960s, where every group was given more or less equal footing by planners, may be over. Rather than attempting to provide services to all neighborhoods, there appears to be a stress on neighborhoods with the strongest prospect of making it. Even in the Cleveland program, with its concern for groups with little existing strengths, Krumholz made it clear that planners should deal with problems in neighborhoods where there was some likelihood of success.

Eplan mentioned an interesting aspect of this innovative practice having to do with staff resources. Having identified 190 neighborhoods in Atlanta, it was obvious that substantial staff services were not available for all. Atlanta's planners developed *Neighborhood Planning Units* that were composed of six to eight neighborhoods of roughly 22,000 to 23,000 people—quite a change from the traditional use of the term *Neighborhood Planning Unit,* which has been that population served by an elementary school. The planners then provided technical services and planning guidelines to these units. Yet Eplan also noted that eleven of these 190 neighborhoods were given a special classification called "reviving," which allowed for extraordinary staff services and support. This seems like a reasonable way of dealing with a problematic situation that is common to all cities in this group.

While we consider this movement to a neighborhood level of planning as the most significant of city planning practices described as innovations by the city planners we talked with, there is an interesting variation having to do with public-private joint ventures for downtown development. We relate this innovation to neighborhoods because, in all of the cases discussed, there was a linkage between new downtown developments and neighborhood stability. The argument was that a revitalized central city was essential to provide the jobs, services, and cultural activities for these redeveloping neighborhoods. The highest level of innovation for public-private joint ventures for downtown redevelopment appeared to us to be in Kansas City, Indianapolis, and Milwaukee. The city planners in these cities stressed that the central city must help developers with all stages of the development process from planning to occupancy. Vitt seemed to say that in the strongest terms. He believed that the city planner should get in on the initial planning and work with the developer until completion. Carroll was not willing to go quite that far, but instead he believed that developments that were related to the plans for the downtown area of Indianapolis should receive technical assistance and support from the planners—including assistance with arranging financing. Drew seemed to imply that planning in Milwaukee would help with new downtown developments that had a favorable rate of return for the city and expanded the set of developments to include industrial buildings—this was not really mentioned in other cities.

An interesting variation on this innovation was a mild dissent from two planners. Krumholz discussed the example of a major downtown development that he opposed—something that would have been heresy elsewhere. His argument was that the benefits would accrue to only a select group of privileged residents and not the people he believed needed such benefits. Costs would accrue to the residents of Cleveland. He lost the struggle. On another track, Eplan seemed to be saying that the recent development burst, in which almost any kind of new downtown project was encouraged, hurt Atlanta. He seemed to imply that the planners in Atlanta were going to be more careful about which developments were essential and consistent with planning goals. This would seem to indicate that some developments might not be consistent and conceivably could be opposed. This is no doubt due to a serious overbuilding for several categories of uses in Atlanta in recent years. Nonetheless, we consider it to be an innovation that almost all of these planners consider public-private joint ventures for downtown developments to be a vital part of neighborhood rejuvenation.

The recent fiscal crisis of New York City, and several other cities for that matter, seemed to alert city planners to a renewed concern for relating fiscal matters to planning. This is an innovation from the recent practice that was an outgrowth of the planning-programming-budgeting approach of the 1960s. The difference appears to be one of lessened expectations and induced fiscal conservatism. In Milwaukee, a city well-known for fiscal tightness, Drew openly discussed the need for fiscal balance in planning, which is another way of saying that development must result in a favorable flow of revenues to the city. Spaid made it clear that neighborhood planning was the key to influencing budgetary decision-making. Carroll seems convinced that making an input to the budget-making process is the most critical achievement for planners. Eplan discussed the innovation in terms of the procedural basis established by the new Charter and how Neighborhood Planning Units were to be related to this process. The point seems clear to us. Most planners have come to realize that the most important aspect of city planning is that which involves spending money. The planners will have to play a major role in that process if implementation is to be taken seriously.

Why should we consider this an innovation? Planners have long been concerned with capital improvements programming and intricate processes for linking planning to budgeting. The reason we see this as an innovation is because many of the planners in our series described how they are going a step further in the process. That step entails an extension into the political arena in which such budgetary decisions are made. By that we mean that the planner may be going beyond the traditional role of recommending a given project to that of fostering and championing a project where it is part of a major planning goal. There also are signs that some of this budget-making influence is being exerted in the operating budget as well. The capital budget is critical, but there are many aspects of the operating budget of a city that dramatically affect planning programs. By showing increased concern for these operating expenditures, we believe that planners are being innovative in attempting to influence the short-term as well as the long-term effects of planning.

Two innovations that we can note are not necessarily widespread. The planning program discussed by Krumholz for Cleveland obviously is innovative. It differs markedly from previous efforts in the social planning context that were variously described as planning for equal opportunity, advocacy, or social equality. The Cleveland program appears to be an inclusion of all of these, yet it is distinctive. The distinction is that it works. City planning that is inherently directed toward meeting the needs of city residents who have little choice or chance appears to have attained an institutionalized base in Cleveland. To direct the major part of a city planning program in this direction certainly is innovative.

On the other hand, the sort of growth-control orientation of the Portland planning program is innovative in terms of Bonner's ideology. Far more than just a "keep-the-newcomers-out" sort of mentality, Bonner is concerned with a reordering of quality of life alternatives. This extends to housing, commuting, recreation, and working, and takes a direction that conspicuously contradicts mass consumption. It is a planning program that takes on an aura of a new breed of individualism linked with environmental and philosophical beliefs. This is innovative in our judgment because it is most uncommon elsewhere. It is all the more interesting because it has received a basis of political support.

We view these items as innovations in city planning practice that derive from the conditions and perspectives of the cities involved. They may have emanated more from trends and patterns of society rather than theory. They may be based in movements that were created by a variety of inherent factors that cried out for some set of responses by city governments. We believe it most interesting that in the cases discussed it was the city planner, rather than another city official, that was instrumental in providing the demanded responses.

SUCCESSES AND FAILURES

We originally sought to determine the successes and failures of city planning in the cities under consideration. We believed such discussion would provide a basis for comparative evaluation. As may be apparent to the reader, it is still most difficult to get city planners or other public officials to discuss and articulate their failures. Most of the city planners we talked with went into some detail about their successes and innovations, yet there was hesitancy to volunteer examples of failures. Even when we probed for information on examples of failures, we could receive only cautious explanations.

Two observations may be relevant to this finding. The first is that all of the planners in this series think of planning as a continuous, cooperative, and comprehensive process. This means that it does not really have an end-state and, hence, makes evaluation always premature. The process itself can be evaluated and revised, but the failure of planning cannot be demonstrated. To demonstrate that there have been failures in planning, one would have to adopt a strict view that it is disjointed, incremental, and temporal, to the extent that it has a beginning and an end. This leads to the second observation, which is that most of these planners concede that there is incrementalism in planning but only to the extent that it is directed to a comprehensive set of goals. This means that a failure somewhere along the line is not a drastic one, since the process still continues toward the comprehensive goals. In other words, nothing short of the Chicago Fire could be regarded as a major failure of planning in Chicago. This may be overstating the rationale, but it is indicative of the very difficult problem that exists for evaluation of planning.

The one possible exception to some of the above statements was our talk with Krumholz. He seemed to delight in discussing his losing battles with private power companies and major downtown developers. He was not at all shy to mention his overwhelming defeat by the Council, nor that there were numerous demands generated for his head. He retained both his head and his job because he was approaching planning in a quite different manner than were the other planners in this series. He does not see planning in the continuous, cooperative, and comprehensive mold; instead he sees it as having a clear mission and cause—perhaps this is a socialized form of incrementalism. Thus he failed miserably when he attempted to stop a development on Lake Erie because it did not provide land for a public beach and park. He was swamped when he tried to stop a major downtown tower development because he did not think there was enough benefit accruing to the unemployed in Cleveland. He was almost run out of town when he tried to influence the way that the police patrolled neighborhoods. Each of these failures may be a basis for in-depth case study in itself. Krumholz is no Don Quixote, however. He entered each of these forays with a clear mission linked to attainment of the overriding goal set for his planning agency. He was not tilting at windmills, but he was going after what he perceived to be the bad guys. Even though he lost a few battles, he survived to fight others. In that sense, Krumholz has not failed.

Having said all of that, we now would like to pose the proposition that there was one particular group of statements in all of the talks that leads us to believe that there is a major failure in city planning. That failure has to do with the movement to develop a true form of citizen participation in city planning. All of the planners we talked to said something to the effect that, "I favor citizen participation, *but* . . . ." That is indicative that all is not well with citizen involvement—perhaps it may indicate a major failure.

Certainly much of this implied failure must be related to the silly concept of "maximum feasible participation" enunciated during the War on Poverty of the 1960s, a concept so devilishly dissected by Daniel Patrick Moynihan in his book *Maximum Feasible Misunderstanding.* Apparently there is still a source of resentment toward the massive, time-consuming, and futile efforts to restore the New England Town Meeting approach to planning that was created by the War on Poverty.

All of the planners we talked with showed a certain amount of skepticism and bitterness toward such experiments. Consider that Krumholz, apparently the most socially oriented of the group, said that "if we went ahead and followed through on some of the things that these neighborhood groups were saying, ... we would be in the same position as were the 'Good Germans'." In other words, Krumholz believes in citizen participation as much as anyone, but he has seen federally imposed experiments at massive citizen involvement in planning turn toward "antidemocratic" stands. This means to us that we have failed in this area because we only have a romantic notion from simpler times to use as a basis for citizen participation in a complex, large urban society. (We should also point out that the much admired New England Town Meeting also burned several "witches.")

Another indicator of the failure with citizen participation was the common statement to the effect that citizens know what is best for them, but it is the politician that represents them and makes the decision. This common attitude among the planners has several interesting overtones and implications. Consider that Eplan has said that citizen participation has "enormously improved the quality and relevance of the decisions we are making." Yet he cited problems, such as poor organization, reacting without proper information, forming opinions that are highly parochial and personal, and banding together when threatened yet being apathetic when not. This leads him to conclude that even with the impressive requirement for citizen participation in the Atlanta City Charter, "rarely should citizens' voices be a substitute for that of the elected official who is ultimately responsible for the welfare of the citizens and city."

How can this statement, which was echoed in many different ways by many in this group of planners, be used as an indicator of success or failure? Bonner seemed to be saying that the real trick to city planning was developing a set of services by which people could help themselves. He somewhat eschewed the conventional wisdom of massive citizen participation on the basis that it led to neutralization in setting goals and a tendency to demand more than local government could deliver. Spaid took a somewhat different track but one that was related to what Eplan and Bonner had discussed. Spaid argued that city planning had to have broad-based citizen support, usually emanating from the neighborhoods, as well as the backing of the elected officials. He went a step further, however, and argued that city planners should provide the leadership for both citizen and elected official backing of planning. He meant by that to describe a role whereby the planner expounds a set of goals and values rather than acting as a secretariat for a debate forum. This leadership role for planners is another of those aspects of practice that has received little if any consideration in the study of the profession.

The question of citizen participation in the city planning process is a dilemma. Every one of the planners we talked with gave whole-hearted support to citizen participation in the planning process. Yet to some that meant holding a public hearing on plans and little else. To others it meant dealing more directly with elected officials, since they are responsible for the citizens' welfare. Still others said that simply recommending what citizens' groups wanted would lead to undemocratic conditions in cities. The dilemma that arises is that since every one supports it, why does everyone think that it has not and will not work—if applied in literal terms?

We suspect that the problem is one that has deep roots in American Democracy and has reached unprecedented complexity in modern cities. The fragment of a solution would seem to lie in some form of neighborhood or neighborhood district participation by citizens at that level of planning. Yet we have heard that it is precisely at this level where the most parochial forms of involvement appear. Perhaps the problem is related to the information that citizens are supplied with or that they demand to know about. In other words, are professional planners asking too much of citizens, and are citizens demanding to become involved in too much detail of city planning—the problem of not seeing the forest for the trees?

The apparent failure of citizen participation is linked as well to our inadequate knowledge and reluctance to use new technologies. It would appear that our only operating model of participation, whether through group meetings or public hearings, is the New England Town Meeting. This is a comfortable and nostalgic model despite its obvious anachronistic character. Why are we not using more of the techniques of technology and survey research to create public involvement and awareness? It certainly can be proven, for example, that a well-designed random sample polling of citizens, let us say by mail or telephone, would offer a more accurate representation of attitudes and opinions in a statistical frame of reference. If it is the face-to-face meeting that is essential, and if the surveys and polls are mistrusted, why not employ nominal group techniques or other forms of group dynamics that will likely lead to more useful information? We suspect that there is an inherent strain of know-nothingism and distrust of government that is so ingrained in the American character that such plausible resolutions of the problem will always be criticized.

We believe that the set of implications that we have generalized indicates a failure of citizen participation that merits serious study. The consequences of inattention to the problem could be dangerous, whether they lead to refusal to make citizen involvement an important endeavor or to a return to the mass meeting mentality of the 1960s. While nobody in the city planning profession would dare challenge the platitude that "planning is for the people," we have seen sufficient evidence that indicates the likelihood that we do not really know what the people want.

LESSONS LEARNED

In attempting to draw conclusions from these talks with planners that might be deemed as lessons to be learned for other cities, our first inclination was a taxonomy. The order and specification that could be offered by a taxonomy of city planning practice was appealing as a way of generalization. Could we say that the approach described by Bonner was that of Growth Control Planning? Could we describe the work of Krumholz as Social Planning? Could we say that Vitt was dealing in Pro-Development Planning or that Eplan was combining Neighborhood Planning with Planning and Budgeting? As tempting as that was to our academic minds, it obviously was fruitless and superficial. We have learned that city planning practice is much too complex to place in neat boxes labeled with catchy adjectives.

If there are lessons to be learned from these talks with city planners, we suspect that they are lessons about people and their needs in large central cities. These needs may vary from city to city and depend greatly upon the social and economic conditions and characteristics of each city. Nonetheless, each city has definable needs, whether they be keeping things about the way they are or trying desperately against seemingly hopeless odds to make things better. The lessons to be learned are centered upon such realities.

We think the best way to describe the major lesson to be learned is to use a political analogy—constituencies must be served. We believe that the success of city planners will depend upon the leadership and technical skills they have to offer to meet the expressed or implied—perhaps induced— demands of constituencies that they serve. This is related to our previous discussion of problems of citizen participation as well as to the politicization of planning practice at the higher levels of the organizational hierarchy. Yet it is more relevant to the manner in which professional planners in leadership positions can define a constituency for planning services and satisfy that constituency.

The constituency may be different in various cities. Our best example of that is the implicit metamorphosis that Bonner went through as he moved on from the low-income, depressed constituency of Cleveland to the middle-income, intellectual constituency of Portland. The man did not change, either in planning philosophy or in skills, but the applications had to be modified. Rather than concentrating upon people with least choice, Bonner was to deal with people demanding implementation of their choice for lifestyle preference. Bonner and his Mayor did not create or change that constituency—they identified it and served its needs. We believe that this most difficult of planning problems may lie at the crux of success in implementation.

There is further evidence of differing constituencies for planning among cities. The constituency being served in Kansas City may include those with least choice (as in Cleveland) and those with growth control on their minds (as in Portland), but the major constituency being served would seem to be a development and growth-oriented group. We know that includes developers, realtors, politicians, and downtown businesspeople in Kansas City, but we suspect it includes a sizeable chunk of the citizenry as well. The Indianapolis and Milwaukee planning programs appear to be moving in similar directions, yet each has decidedly varying components of the constituency. The constituency that Eplan and Spaid are interested in serving appears to be largely definable by neighborhood groupings; they believe that these constituencies can sum to a total for citywide planning. While Carroll is concerned with neighborhoods as well, he must consider his constituency to be the urban region that was created by political changes. Neither Spaid nor Eplan, however, saw much value in regional constituencies in their planning programs, probably because the regional planning programs in their areas are distinct and themselves serve a regional constituency.

We believe that a city planner must learn to identify these constituencies and serve them well for both success and survival. This is most complicated in that there are many constituencies to choose from in most cities. This means that a city planner must be aware of the political, as well as social and economic, ramifications of constituencies. In this line of thinking, it may appear to be safer to deal with neighborhoods, since they usually have some political representation. This would apply as well to consolidated cities, since at-large representation has a set of defined group interests that are aligned with elected leaders. The most dangerous would appear to be a socioeconomic constituency, either poor or rich, since these groups go through various cycles of political power and significance.

There also is the mysterious lesson related to leadership that city planners must learn. Some of the planners we talked with believed that constituencies must be represented by their own leadership and that their fortunes rise and fall with such people. Other planners believed that politically selected leaders offer a more direct tie-in with the way the decisions for planning are made. There also is a group of planners that is inclined to provide such leadership where none exists. This is a vexing and risky lesson. If mistakes are made—the wrong people consulted and the correct people ignored—the consequences could be disastrous for the planner in charge. While we have had suggestions on how to seek the truth—ranging from Spaid's nocturnal tavern discussions to Eplan's neighborhood visitations—we do not seem to have a clear process for identifying the leadership of constituencies to be served.

Another lesson we have learned is one in which the literature is virtually silent, yet our intuition leads us to suspect its presence. The individual as city planner is a major factor of success. Such individual traits as leadership, values, morality, commitment, and work ethic are strong factors for the implementation of planning. We may sometimes call this *personality*, even though planners at conferences and in writing like to eschew personality as a factor, or we may say more acceptably to a professional audience that a planner needs a good *style*. Whatever term we may use, we still mean to say that individuals may do things in different ways, even to achieve the same ends, and that in itself has a great deal to do with success and survival. Spaid likes to describe this as professionalism and imply that it sets a standard for behavior by planners. Krumholz prefers to describe this phenomenon as social concern and the sacrifice that goes along with it. Yet we suspect that both are talking about the same concept of personality and style. Those are things that are not acquired in school, but they are behavioral and belief patterns developed over a period of time through experience and interaction with an environment and groups within it. We fully suspect that personality and style have much to do with the selection process of city planners for given positions in various cities. We suspect that when the selection process is invalid or conditions change, the political powers that make such selections will correct the mistake. Thus, we can see that planning directorships may change—either slowly or rapidly— but that is not such a bad thing if we are more concerned with implementation and action than with the technical process of planning.

Perhaps it is solely a factor of the group of planners we talked with, but we were impressed by the inherent optimism in each of these people. There were no real pessimists in the group. While Krumholz may have seemed a little angry, he was perhaps the most optimistic of the lot—given that he may well have the toughest job. This optimism was not typified as the crusading zealot but as the operational pragmatist. They tend to pooh-pooh planning theory that was promulgated by cynical intellects and profess an optimistic pragmatism based on the need to bring complex problems down to a human level and solve those problems in the simplest and most direct manner. This means to us that the planning profession must be quite healthy, peopled by optimists, and concerned with an indigenous approach to making the quality of life better for people living in large cities. This is not always well understood.

Our final lesson to be learned is perhaps the most important one of all. We have talked with seven of the leading city planners in large central cities of the United States. We have learned from them that city planning is being taken very seriously by both politicians and constituents. That has not always been the case in this country. Planning has always been considered with some degree of skepticism and suspicion. Be that as it may, it appears that planning has gone beyond the institution-building stage and may be entering a halycon period of acceptance. This gives pause for encouragement and a sense of controlling our destinies. It leads us to suspect that people in these United States have come to realize that their present needs and future aspirations may be more possible due to a planning process and way of thinking that is very human in perspective.

APPENDIX

TABLE 1
City Population, 1900 - 1970

	1900	1910	1920	1930	1940	1950	1960	1970
Atlanta								
Population	89,872	154,839	200,616	270,366	302,288	331,314	487,455	496,973
Percent Change		72.3	29.6	34.8	11.8	9.6	47.1	2.0
Cleveland								
Population	381,768	560,663	796,841	900,429	878,336	914,808	876,050	750,903
Percent Change		46.9	42.1	13.0	-2.5	4.2	-4.2	-14.3
Indianapolis								
Population	169,164	233,650	314,194	364,161	386,972	427,173	476,258	744,624
Percent Change		38.1	34.5	15.9	6.3	10.4	11.5	56.3
Kansas City								
Population	163,752	248,381	324,410	399,746	399,178	456,622	475,539	507,087
Percent Change		51.7	30.6	23.2	-0.1	14.4	4.1	6.6
Milwaukee								
Population	285,315	373,857	457,147	578,249	587,482	637,392	741,324	717,099
Percent Change		31.0	22.3	26.5	1.6	8.5	16.3	-3.3
Portland								
Population	90,426	207,214	258,288	301,815	305,394	373,628	372,676	382,619
Percent Change		129.2	24.6	16.9	1.2	22.3	-0.3	2.7
St. Paul								
Population	163,065	214,744	234,698	271,606	287,736	311,349	313,411	309,980
Percent Change		31.7	9.3	15.7	5.9	3.2	0.7	-1.1

Source: U.S. Bureau of the Census. Census of Population: 1970. **Vol. 1, Characteristics of the Population** (Washington, D.C.: U.S.G.P.O., 1973).

Appendix

Population and Population Changes, Standard Metropolitan Statistical Areas, 1960-1973

	1960	1970	1973	Percent Change 60-70	Percent Change 70-73	Annual Change 60-70	Annual Change 70-73
Atlanta							
SMSA	1,017,188	1,390,164	1,747,987	36.7	25.7	3.7	8.6
Central City	487,455	496,973		2.0		.2	
Outside Central City	529,733	893,191		68.6		6.9	
Cleveland							
SMSA	1,909,483	2,064,194	2,006,371	8.1	-2.8	.8	-.9
Central City	876,050	750,903		-14.3		-1.4	
Outside Central City	1,033,433	1,313,291		27.1		2.7	
Indianapolis							
SMSA	944,475	1,109,882	1,136,598	17.5	2.4	1.8	0.2
Central City*	476,258	743,155		56.0		5.6	
Outside Central City	468,217	366,727		-21.7		-2.2	
Kansas City							
SMSA	1,092,545	1,253,916	1,298,849	14.8	3.6	1.5	1.2
Central City*	475,539	501,859		5.5		.6	
Outside Central City	617,006	752,057		21.9		2.2	
Milwaukee							
SMSA	1,278,850	1,403,887	1,416,773	9.8	.1	1.0	0.0
Central City	741,324	717,372		3.2		0.3	
Outside Central City	537,526	686,515		27.8		2.9	
Portland							
SMSA	821,897	1,009,129	1,062,451	22.8	5.3	2.3	1.8
Central City	372,676	382,619		2.7		.3	
Outside Central City	449,221	626,510		39.5		4.0	
St. Paul (Minn.-St. Paul)							
SMSA	1,482,030	1,813,647	1,999,753	22.4	10.3	2.2	3.4
Central City	796,287	744,380		-6.5		-.7	
Outside Central City	685,747	1,069,267		55.9		5.6	

*Urban Part Only

Sources: 1960 and 1970: U.S. Bureau of the Census. Census of Population: 1970. Vol. 1, Characteristics of the Population (Washington, D.C.: U.S.G.P.O., 1973).

1973 Estimates: U.S. Bureau of the Census, Local Government Finances in Selected Metropolitan Areas and Large Counties, 1974-75, Series 6F-75, No. 6 (Washington, D.C.: U.S.G.P.O., 1976).

TABLE 3

Components of Population Change, 1960-1970

		Natural Increase (Decrease)	Net Migration	Net Migration as Percent of 1960 Population
Atlanta	(City)	59,285	-50,319	(-) 10.3
	(SMSA)	172,918	200,058	19.7
Cleveland	(City)	83,995	-209,166	(-) 23.9
	(SMSA)	199,677	-44,966	(-) 2.4
Indianapolis	(City)	96,745	-2,013	.0
	(SMSA)	129,668	37,030	3.9
Kansas City	(City)	47,346	-15,798	(-) 3.3
	(SMSA)	134,477	29,305	2.7
Milwaukee	(City)	81,125	-105,077	(-) 14.2
	(SMSA)	163,364	-38,327	(-) 3.0
Portland	(City)	12,847	-4,968	(-) 1.3
	(SMSA)	68,729	116,504	14.2
St. Paul	(City)	34,854	-38,437	(-) 4.8
Minneapolis-St. Paul	(SMSA)	232,330	99,421	6.7

Sources: Barton-Aschman, Inc. **Urban Transportation Fact Book** (American Institute of Planners and Motor Vehicle Manufacturers Association of the U.S., Inc., March, 1974).

U.S. Bureau of the Census. Census of Population: 1970. **Vol. 1, Characteristics of the Population** (Washington, D.C.: U.S.G.P.O., 1973).

TABLE 4

Black Population: 1970

	SMSA			City		
	Total Population	Black Population	Percent Black	Total Population	Black Population	Percent Black
Atlanta	1,390,164	310,632	22	496,973	255,051	51
Cleveland	2,064,194	332,614	16	750,903	287,841	38
Indianapolis	1,109,882	137,335	12	*743,155	*134,040	18
Kansas City	1,253,916	151,127	12	*501,859	*112,004	22
Milwaukee	1,403,688	106,532	08	717,099	105,088	15
Portland	1,009,129	23,284	02	382,619	21,572	06
St. Paul	1,813,647	32,118	02	309,980	10,930	04

*Urban Part

Source: Table 23. Race by Sex, for Areas and Places: 1970. U.S. Bureau of the Census. Census of Population: 1970. **Vol. 1, Characteristics of the Population** (Washington, D.C.: U.S.G.P.O., 1973).

Appendix

TABLE 5
Nativity (SMSA Population): 1970

	Percent Foreign-Born	Percent Native of Foreign or Mixed Parentage
Atlanta	1.2	2.7
Cleveland	6.9	18.6
Indianapolis	1.2	3.7
Kansas City	1.6	5.9
Milwaukee	4.5	16.1
Portland	4.1	12.9
Minneapolis-St. Paul	3.0	4.7

Source: Table 40. Summary of Social Characteristics: 1970. U.S. Bureau of the Census. Census of Population: 1970. **Vol. 1, Characteristics of the Population,** (Washington, D.C.: U.S.G.P.O., 1973).

TABLE 6
Employed Persons, 1970

		Percent in Manufacturing	Percent in White-Collar Occupations
Atlanta	(City)	16.7	49.7
	(SMSA)	19.7	57.9
Cleveland	(City)	37.5	36.6
	(SMSA)	35.4	50.1
Indianapolis	(City)	27.9	52.1
	(SMSA)	29.2	50.0
Kansas City	(City)	21.2	52.4
	(SMSA)	22.7	53.7
Milwaukee	(City)	34.8	45.0
	(SMSA)	35.1	48.9
Portland	(City)	17.4	55.1
	(SMSA)	21.0	52.8
St. Paul	(City)	25.1	53.7
Minneapolis-St. Paul	(SMSA)	24.7	56.7

Source: Table 41. Summary of Economic Characteristics: 1970. U.S. Bureau of the Census. Census of Population: 1970. **Vol. 1, Characteristics of the Population,** (Washington, D.C.: U.S.G.P.O., 1973).

TABLE 7
Civilian Labor Force, Percent Unemployed: 1970

		Total Population	Black Population
Atlanta	(City)	4.0	4.0
	(SMSA)	3.0	3.8
Cleveland	(City)	5.2	7.7
	(SMSA)	3.5	7.2
Indianapolis	(City)	4.3	8.5
	(SMSA)	3.9	8.4
Kansas City	(City)	3.8	6.1
	(SMSA)	3.3	6.5
Milwaukee	(City)	4.1	8.3
	(SMSA)	3.5	8.2
Portland	(City)	6.6	11.9
	(SMSA)	6.1	11.4
St. Paul	(City)	3.6	7.6
Minneapolis-St. Paul	(SMSA)	3.2	6.4

Sources: Table 41. Summary of Economic Characteristics: 1970. U.S. Bureau of the Census. Census of Population: 1970. **Vol. 1, Characteristics of the Population** (Washington, D.C.: U.S.G.P.O., 1973).

Table 92. Employment Characteristics of the Negro Population for Areas and Places: 1970. U.S. Bureau of the Census. Census of Population: 1970. **Vol. 1, Characteristics of the Population** (Washington, D.C.: U.S.G.P.O., 1973).

TABLE 8
Median Income, 1970 (families)

Atlanta	(City)	8,399
	(SMSA)	10,695
Cleveland	(City)	9,107
	(SMSA)	10,257
Indianapolis	(City)	10,754
	(SMSA)	10,754
Kansas City	(City)	9,910
	(SMSA)	10,568
Milwaukee	(City)	10,262
	(SMSA)	11,338
Portland	(City)	9,799
	(SMSA)	10,463
St. Paul	(City)	10,544
Minneapolis-St. Paul	(SMSA)	11,682

Source: Table 41. Summary of Economic Characteristics: 1970. U.S. Bureau of the Census. Census of Population: 1970. **Vol. 1, Characteristics of the Population** (Washington, D.C.: U.S.G.P.O., 1973).

Appendix

TABLE 9
Family Income, 1970

		Families with Income Below Poverty Level	Families with Income $15,000+
Atlanta	(City)	15.9	18.9
	(SMSA)	9.1	26.1
Cleveland	(City)	13.9	15.3
	(SMSA)	6.9	28.2
Indianapolis	(City)	7.1	24.9
	(SMSA)	6.5	23.9
Kansas City	(City)	8.9	20.2
	(SMSA)	6.9	23.0
Milwaukee	(City)	8.1	19.2
	(SMSA)	5.7	26.1
Portland	(City)	8.1	20.5
	(SMSA)	6.9	22.2
St. Paul	(City)	6.4	22.2
Minneapolis-St. Paul	(SMSA)	4.6	28.6

Source: Table 41. Summary of Economic Characteristics: 1970. U.S. Bureau of the Census. Census of Population: 1970. **Vol. 1, Characteristics of the Population** (Washington, D.C.: U.S.G.P.O., 1973).

TABLE 10
Median House Value (owner occupied), 1970

Atlanta	(City)	17,000
	(SMSA)	19,800
Cleveland	(City)	16,700
	(SMSA)	22,800
Indianapolis	(City)	14,800
	(SMSA)	14,900
Kansas City	(City)	14,100
	(SMSA)	15,900
Milwaukee	(City)	18,200
	(SMSA)	21,500
Portland	(City)	14,400
	(SMSA)	16,700
St. Paul	(City)	18,600
St. Paul-Minneapolis	(SMSA)	21,500

Source: Table 1. Summary Characteristics for Areas and Places: 1970. U.S. Bureau of the Census. Census of Housing: 1970. **Vol. 1, Housing Characteristics for States, Cities, and Counties** (Washington, D.C.: U.S.G.P.O., 1972).

TABLE 11
Owner Occupancy: 1970

		Percent Owner Occupied
Atlanta	(City)	41.2
	(SMSA)	57.5
Cleveland	(City)	46.1
	(SMSA)	62.4
Indianapolis*	(City)	61.3
	(SMSA)	65.4
Kansas City*	(City)	58.0
	(SMSA)	65.7
Milwaukee	(City)	47.3
	(SMSA)	59.8
Portland	(City)	56.5
	(SMSA)	65.0
St. Paul	(City)	56.3
Minneapolis-St. Paul	(SMSA)	65.2

*Urban Part

Source: Table 8. Occupancy, Plumbing, and Structural Characteristics for Areas and Places: 1970. **Vol. 1, Housing Characteristics for States, Cities, and Counties** (Washington, D.C.: U.S.G.P.O., 1972).

TABLE 12
Over Crowding (1.01 or more persons per room): 1970

Atlanta	(City)	11.0
	(SMSA)	7.6
Cleveland	(City)	7.4
	(SMSA)	5.5
Indianapolis	(City)	8.2
	(SMSA)	7.9
Kansas City	(City)	6.5
	(SMSA)	6.2
Milwaukee	(City)	7.3
	(SMSA)	6.8
Portland	(City)	3.5
	(SMSA)	4.4
St. Paul	(City)	6.1
Minneapolis-St. Paul	(SMSA)	6.7

Source: Table 1. Summary Characteristics for Areas and Places: 1970. U.S. Bureau of the Census. Census of Housing: 1970. Vol. 1, **Housing Characteristics for States, Cities, and Counties** (Washington, D.C.: U.S.G.P.O., 1972).

Appendix

TABLE 13
Year Structures Built (all units), 1970

		Total	1939 or Earlier	Percent	1969-1970	Percent
Atlanta	(City)	170,863	51,741	30.2	4,657	2.7
	(SMSA)	450,202	81,318	18.1	29,062	6.5
Cleveland	(City)	264,156	193,763	73.4	1,351	0.1
	(SMSA)	676,085	310,594	45.9	14,120	2.1
Indianapolis*	(City)	251,821	100,092	39.7	9.921	3.9
	(SMSA)	363,591	143,515	38.9	15,630	4.2
Kansas City*	(City)	190,700	98,110	51.4	4,129	2.2
	(SMSA)	435,821	164,690	37.8	13,771	3.2
Milwaukee	(City)	245,995	135,394	55.0	3,325	1.4
	(SMSA)	446,516	203,565	45.6	10,152	2.3
Portland	(City)	151,839	86,868	57.2	2,203	1.5
	(SMSA)	357,351	132,362	37.0	15,694	4.4
St. Paul	(City)	107,694	67,234	62.4	2,687	2.5
Minneapolis-St. Paul	(SMSA)	574,826	224,960	39.1	27,753	4.8

*Urban part

Source: Table 43. Structural and Plumbing Characteristics for Areas and Places: 1970. U.S. Bureau of the Census. Census of Housing, 1970. **Vol. 1, Housing Characteristics for States, Cities, and Counties** (Washington, D.C.: U.S.G.P.O., 1972).

TABLE 14
Percent Who Completed Four Years of High School or More: 1970
(persons 25 years old and over)

Atlanta	(City)	46.5
	(SMSA)	53.4
Cleveland	(City)	37.4
	(SMSA)	54.6
Indianapolis	(City)	54.8
	(SMSA)	56.0
Kansas City	(City)	55.9
	(SMSA)	60.1
Milwaukee	(City)	49.2
	(SMSA)	56.7
Portland	(City)	60.4
	(SMSA)	62.9
St. Paul	(City)	57.2
Minneapolis-St. Paul	(SMSA)	66.1

Source: Table 40. Summary of Social Characteristics: 1970. U.S. Bureau of the Census. Census of Population: 1970. **Vol. 1, Characteristics of the Population** (Washington, D.C.: U.S.G.P.O., 1973).

TABLE 15
Land Area and Density, 1970

		Land Area (square miles)	Density (persons per square mile)
Atlanta	(City)	131.5	3,779
	(Urbanized Area Outside of Central City)		3,227
	(SMSA)	1728	804
Cleveland	(City)	25.9	9,893
	(Urbanized Area Outside of Central City)		2,120
	(SMSA)	1519	1,359
Indianapolis	(City)	351.7	2,113
	(Urbanized Area Outside of Central City)		2,641
	(SMSA)	3080	361
Kansas City	(City)[a]	238.9	2,010
	(Urbanized Area Outside of Central City)		2,359
	(SMSA)	2768	454
Milwaukee	(City)	95.0	7,548
	(Urbanized Area Outside of Central City)		1,481
	(SMSA)	1456	964
Portland	(City)	89.1	4,294
	(Urbanized Area Outside of Central City)		2,489
	(SMSA)	3650	276
St. Paul	(City)[b]	105.6	7,049
Minneapolis-St. Paul	(Urbanized Area Outside of Central City)		1,559
	(SMSA)	2108	860

[a] Includes both Kansas City, Kansas, and Kansas City, Missouri
[b] Includes both Minneapolis and St. Paul

Source: Tables 1-3 (Land Areas, 1970) and 1-4 (Population Density, 1970), Barton-Aschman, Inc. **Urban Transportation Fact Book** (American Institute of Planners and Motor Vehicle Manufacturers Association of the U.S., Inc., March, 1974).

Appendix [219]

TABLE 16
Work Trips, by Transportation Mode, 1970

		Auto	Public Transit
Atlanta	(City)	71	21
	(SMSA Ring)	92	3
Cleveland	(City)	68	22
	(SMSA Ring)	84	9
Indianapolis	(City)	84	8
	(SMSA Ring)	87	1
Kansas City[a]	(City)	81	10
	(SMSA Ring)	91	2
Milwaukee	(City)	70	19
	(SMSA Ring)	84	4
Portland	(City)	76	11
	(SMSA Ring)	88	3
Minneapolis-St. Paul (City)[b]		69	17
Minneapolis-St. Paul (SMSA Ring)		89	3

[a] Includes both Kansas City, Kansas, and Kansas City, Missouri
[b] Includes both Minneapolis and St. Paul

Source: Tables 1-15, How People Travel to Work, 1970, Barton-Aschman, Inc. **Urban Transportation Fact Book** (American Institute of Planners and Motor Vehicle Manufacturers Association of the U.S., Inc., March, 1974).

TABLE 17
Percent Change in Workplace Location, 1960-1970
(for SMSA residents)

	SMSA Total	In Central City	Outside Central City	Outside SMSA
Atlanta	46	13	124	11
Cleveland	12	-15	83	124
Indianapolis	58	31	201	130
Kansas City	26	-4	56	19
Milwaukee	19	-11	110	66
Portland	25	8	57	53
Minneapolis-St. Paul	30	0	25	26

Source: Table 1-10 (Percent Change in Workplace Location, 1960-1970), Barton-Aschman, Inc. **Urban Transportation Fact Book** (American Institute of Planners and Motor Vehicle Manufacturers Association of the U.S., Inc., March, 1974).

TABLE 18
Gain-Loss Position under the Community Development Block Grant Program Projection for Fiscal 1980

	Folded-in Programs (thousands of dollars)	Projected CDBG (thousands of dollars)	Percent Gain-Loss
Atlanta	18,780	11,207	-40.3
Cleveland	16,092	14,250	-11.4
Indianapolis	13,929	10,767	-22.7
Kansas City	17,859	8,048	-54.9
Milwaukee	13,383	10,939	-18.3
Portland	8,760	5,483	-37.4
St. Paul	NA	NA	NA

Source: Richard P. Nathan, et al., **Block Grants for Community Development** (Washington, D.C.: U.S. Department of Housing and Urban Development, January, 1977), pp. 158-159.

TABLE 19
Eligibility Index for Community Development Block Grant Funds

Eligible Community	Eligibility Index (percentage above national mean)	Rank (196 eligible cities above the national mean)
E. St. Louis, Ill.[a]	431.86	1
Cleveland, Ohio	291.73	18
Portland, Oregon	141.76	134
Milwaukee, Wisconsin	127.67	149
Kansas City, Missouri	120.58	157
Atlanta, Georgia	118.01	161
St. Paul, Minnesota	116.58	164
Indianapolis[b]		

[a] Included for purposes of comparison
[b] Below the national mean for the eligibility index

Source: Richard P. Nathan, et al., **Block Grants for Community Development** (Washington, D.C.: U.S. Department of Housing and Urban Development, January, 1977), pp. 516-524.

TABLE 20
Rank Order of 65 Largest SMSAs by Quality of Life

SMSA	Rank
Portland	1
Sacramento*	2
Seattle-Everett*	3
Minneapolis-St. Paul	4
Milwaukee	11
Cleveland	24
Kansas City	35
Indianapolis	36
Atlanta	45
Birmingham*	64
Jersey City*	65

*Included for comparative purposes

Source: Ben-Chieh Liu, Quality of Life Indicators in U.S. Metropolitan Areas (New York: Praeger, 1976).

TABLE 21
Quality of Life Indicators: 1970
(Rank 1-65; rating A-E)

	Economic Component		Political Component		Environmental Component		Health & Education Component		Social Component		Overall Quality of Life	
Atlanta	7	A	56	E	52	D	37	D	44	D	45	D
Cleveland	4	A	28	C	60	E	32	C	24	B	24	B
Indianapolis	5	A	41	D	61	E	43	D	38	C	36	C
Kansas City	38	C	50	D	39	C	31	C	12	A	35	C
Milwaukee	16	B	12	A	32	C	19	B	8	A	11	A
Portland	3	A	8	A	11	A	9	A	1	A	1	A
Minneapolis-St. Paul	25	B	9	A	20	B	7	A	9	A	4	A
	Table 1		2		3		4		5		Table 16	

Source: Ben-Chieh Liu, Quality of Life Indicators in U.S. Metropolitan Areas (New York: Praeger, 1976).

INDEX

abandonment, 38, 62, 67, 121
Act 353, 106, 199, 126-217
advocacy, 64-65, 193
advocate planners, 50
Aid to Families with Dependent Children, 61
Alternate Futures Program for Kansas City, 112
Altshuler, Alan, 163, 166
Atlanta Regional Council, 48

Bair, Fred, 169
Barabba, Vincent, 14
Bay Area Rapid Transit District (BARTD), 46-48
Brookings Institution, 28, 112
budgeting, 34-35, 83, 89-90, 97-98
Bureau of Budget Policy and Evaluation, 35
bureaucracy, 54, 55, 58

capital budgeting, 90, 98, 106, 196
capital improvements, 36, 58, 81, 88, 161, 196
Caravelle Commons, 86
Carter, Jimmy, 167
Catanese, A.J., 167
Census Victimization Study, 62
Central Industrial District Association, 114
Citizen participation, 33, 37, 44-45, 49, 73, 172, 199-202
Citizen Participation Ordinance, 50
City Center Square, 127
City Development Department, 104, 106
Clark Freeway, 66

Cleveland City Planning Commission, 63
Cleveland Electric Illuminating, 68, 70, 77
Cleveland Transit System, 66
Community Development Block Grant, 27, 83, 86, 91, 106, 120, 126, 134, 139, 141-142, 156, 161, 167
Community Responsive Transit, 67
comprehensive plan, 33-36, 40, 43, 53, 82, 103, 105, 116, 131, 146, 149-150, 163, 171
Consolidated Cities Act of 1969, 80
Cookingham, Perry, 103
Country Club Plaza Area, 125
Crown Center, 117-118, 128

Davidoff, Paul, 64
decentralization, 25-26
decline, 14, 61
Department of Budget and Planning, 35
Department of Labor, 115
Department of Metropolitan Development, 80
design review, 58
dial-a-bus, 67
Drucker, Peter, 168

Economic Opportunity Act, 20
EDA, 114-115
education, 24-25, 95
employment, 61
energy, 113
equity, 63

Fair Share Housing Allocation Plan, 176
family income, 18-19
Federal Power Commission, 69
FHA, 52
Financial Accounting Management Information System, 98
Fine Arts Advisory Committee, 58
Fiscal Balance, 135, 190
Fortune 500, 60

Georgia Community Development Department, 49
growth control, 188, 197, 204
growth management, 76

Hamilton, Calvin, 59
Hammer-Siler-George Associates, 175
Hedman, Richard, 169
Historic Hill, 161
Hough, 62
Humphrey, Hubert, 160

immigrants, 16
immigration, 12, 22
implementation, 55, 69-70, 81, 84, 86-87, 107, 111, 164, 168, 181
Independent Planning Commission, 186
Industrial Belt, 16
industrial land bank, 142
industrialization, 22
Integrated Grant Administration Program, 82

Jacobs, Jane, 192
Johnson, Tom, 77
Joint Funding Simplification Act, 82

Kaufman, Jerry, 63

land area, 26
land density, 26, 76, 99
land use, 49, 81, 104, 133, 162, 182
Lilly Endowment, 85
Liu, Ben Chieh, 30
Lockerbie Square Historic Preservation Project, 85

Lugar, Richard, 86, 93

Machinery and Equipment Exemption Act, 142
management plan, 109-110
Mandatory Planning Act, 178
Market Square Arena, 84, 100
Martin Luther King Park, 86
mass transportation, 25, 27, 99
master plan, 103, 107, 132
maximum feasible participation, 199
Mauro, John, 59
Mayor Gribbs, 105
Mayor Hudnut, 86
Mayor Latimer, 166
median house value, 20-21
median income, 18-19, 26
Merchants Plaza, 84
Metropolitan Development Department, 97
Missouri Redevelopment Law, 106, 119
Model Cities, 20, 27, 82-83, 85-86, 108
Moffit, Bob, 177
Moynihan, Patrick, 199
Muni-light, 68, 76-77

Nathan, R.D., 28, 29
National Commission on Civil Disorders, 37
Neighborhood Housing Services Program, 121
neighborhood planning, 56, 147, 192
Neighborhood Planning Council, 43
neighborhood planning units, 42-43, 48, 194
Neighborhood Theory, 192
Northeast Ohio Areawide Coordinating Agency, 66

Office of Planning and Development, 146
Open Space Land Grant Program, 86
overcrowding, 22-29
owner occupancy, 21-22

Park Freeway West, 140

Index

Peat, Marwick, and Mitchell, 98
Perk, Ralph, 66
personality, 179-205
Pittsburgh Simulation Model, 59
Planned Industrial Expansion Act, 119, 126
Planning Department, 35
planning process, 45, 55, 107, 132, 169, 181-183, 198
planning theory, 64, 89, 107, 180-182, 187
politicization, 183-186
population, 12-16, 28, 61, 79-80
poverty, 19-20, 28, 61, 64
Prendergast Machine, 104, 123
Preservation Zoning Ordinance, 85
professionalism, 184-185
program budget, 44
PTA, 40
Public Law, 27, 93, 383

Quality of Life Indicators, 30

Real Estate Research Corporation, 121
Reform movements, 186
rehabilitation loans, 27
Reorganization Ordinance, 33-34
Revenue Sharing, 27, 91-92, 120
Riley, James Whitcombe

School of Management and Environmental Affairs, 159
Section 8, 176-177
Section 235, 121
Section 236, 86
701 Program, 97, 139, 162
shipbuilding, 13
social planning, 82, 105, 196, 203
Southeastern Wisconsin Regional Planning Commission, 140
Special Benefit District Law, 126
Sternlieb, G., 31
Stokes, Carl, 66, 70
St. Paul City Planning Commission, 174
strategic planning, 109, 112
strong mayor, 33, 35, 81

style, 205
suburbanization, 20, 26-27
Sunbelt, 11, 24

tax-delinquency, 62
Tennessee Valley Authority, 77
Thompson, Wilbur, 115
town meetings, 199-200
transportation planning, 47, 155
Troost Midtown, 113-114, 116
Twin Cities Metropolitan Council, 160, 176-178

unemployment, 17-18, 83
Unified Government (UniGov) 80, 93-94
University of Minnesota, 159
Urban Economic Development Council, 114
urban renewal, 27, 39, 47, 84, 138, 156, 162, 166
U.S. Department of Housing and Urban Development, 66, 114-115

Wall Street Journal, 166
War on Poverty, 20, 199
weak mayor, 33
Webber, M., 46-47
West Central, 62
work program, 57, 60, 82, 83
workplaces, 26-27
worktrips, 26

zero population growth, 188
zoning, 36, 49, 58, 67, 76, 81, 104, 116, 120, 123-124, 147-149, 163, 182

ABOUT THE EDITORS

ANTHONY JAMES CATANESE has practiced, taught, consulted, and written about urban planning and management. Currently Dean of the School of Architecture and Urban Planning, University of Wisconsin—Milwaukee, he has taught previously at the University of Miami and Georgia Institute of Technology; was a Visiting Professor at the Virginia Polytechnic University and Clark College; and was a Senior Fulbright Professor at the Pontificia Universidad Javeriana, Bogotá, Colombia. His consulting work has spanned the country from New York to Hawaii; and he worked as Senior Planner for the State of New Jersey and Middlesex County, New Jersey. His books include: *Planning and Local Politics: Impossible Dreams; Urban Transportation in South Florida; Systemic Planning: Theory and Application; New Perspectives on Urban Transportation Research;* and *Scientific Methods of Urban Analysis.* He was a member of Jimmy Carter's Urban Policy Task Force during the Presidential Primary and Electoral Campaign.

W. PAUL FARMER has taught in the field of urban and regional planning at the School of Architecture and Urban Planning, University of Wisconsin—Milwaukee, as an Assistant Professor since 1972. Concurrently, he has done consulting work in Wisconsin and New York, and he was a consultant on the development of the Gandhi Krishi Vignana Kendra campus of the University of Agricultural Sciences, Bangalore, India. He was a Planner for the Shreveport Metropolitan Planning Commission in Louisiana. He formerly taught for the Department of City and Regional Planning, Cornell University. He earned the B.A. and B.Arch. from Rice University and the M.R.P. from Cornell University where he is completing requirements for the Ph.D. His affiliations include the American Institute of Planners, where he served on the Task Force on Priorities and Budget, and the Regional Science Association.